Statistical analysis of spatial dispersion A. Rogers

the quadrat method

Monographs in spatial and environmental systems analysis

Series editors R.J.Chorley and D.W.Harvey

1 Entropy in urban and regional modelling A.G.Wilson
2 An introduction to spectral analysis J.N.Rayner
3 Perspectives on resource management T.O'Riordan
4 The mechanics of erosion M.A.Carson
5 Spatial autocorrelation A.D.Cliff and J.K.Ord
6 Statistical analysis of spatial dispersion A.Rogers

p Pion Limited, 207 Brondesbury Park, London NW2 5JN

Statistical analysis of spatial dispersion A. Rogers

the quadrat method

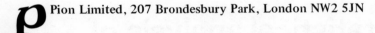 Pion Limited, 207 Brondesbury Park, London NW2 5JN

© 1974 Pion Limited

All rights reserved. No part of this book may be reproduced in any form by photostat microfilm or any other means without written permission from the publishers.

ISBN 0 85086 045 8

Set on IBM 72 Composers by Pion Limited, London.
Printed in Great Britain by J.W.Arrowsmith Limited, Bristol.

Preface

My interest in the potential applicability of quadrat models to spatial analysis was first aroused by some work that I carried out for my dissertation in 1963-1964, when I fitted several such models to data on the spatial dispersion of retail establishments in Stockholm. That experience led me to conclude that further work on the subject could not be fruitfully pursued until very disaggregated and detailed data were to become available on the spatial structure of some city in the United States or abroad. A sabbatical leave spent in Ljubljana, Yugoslavia, in 1967, led to the collection of such data for that city and rekindled my interest in quadrat analysis. Since then, working with the assistance and counsel of three superb students—Norbert Gomar, Juan Martin, and Richard Raquillet—I have pursued my research in this area to the point where the principal findings need to be set out with a greater internal coherence and expositional clarity than is possible in a series of journal articles. Consequently I have collected together my recent papers on quadrat analysis (three of which were coauthored by the above colleagues), added a considerable amount of new and unpublished work, and then redrafted the entire manuscript using a consistent notation throughout.

Although the application focus of this study is retail spatial structure, the statistical techniques used are applicable to a much wider class of spatially distributed phenomena. Thus geographers, economists, marketing specialists, urban planners, demographers, and sociologists will all find that the methods described in this monograph may easily be applied to spatial problems within their particular disciplines. The principal mathematical prerequisites for efficient study of these methods are a course in probability theory at the level of Parzen (1960) or Feller (1957), and a course in mathematical statistics at the level of Lindgren (1962) or Freeman (1963).

I am grateful to Norbert Gomar, Juan Martin, and Richard Raquillet for their kind permission to use material from our joint efforts, and to the publishers and editors of *Geographical Analysis* and *Environment and Planning* for their permission to reproduce parts of the five articles that first appeared in these two journals[†]. To Lidija Podbregar Vasle, formerly with the Town Planning Institute of Slovenia, go my deep thanks for obtaining and mapping the data on Ljubljana's retail and

---

[†] Rogers, A., 1969, "Quadrat analysis of urban dispersion: 1. Theoretical techniques", *Environment and Planning*, 1, 47-80.

Rogers, A., 1969, "Quadrat analysis of urban dispersion: 2. Case studies of urban retail systems", *Environment and Planning*, 1, 155-171.

Rogers, A., Gomar, N., 1969, "Statistical inference in quadrat analysis", *Geographical Analysis*, 1, 370-384.

Rogers, A., Martin, J., 1971, "Quadrat analysis of urban dispersion: 3. Bivariate models", *Environment and Planning*, 3, 433-450.

Rogers, A., Raquillet, R., 1972, "Quadrat analysis of urban dispersion: 4. Spatial sampling", *Environment and Planning*, 4, 331-345.

population spatial structures. Noreen Roberts of the Bay Area Transportation Study staff very kindly provided the data on the spatial distribution of retail establishments in San Francisco. Professor J. B. Douglas of the mathematical statistics department at the University of New South Wales, Australia, and Dr. Jan Hoem of Norway's Central Bureau of Statistics generously read parts of the manuscript, offered several useful suggestions, and identified errors in the exposition. Finally, in writing this monograph, I have drawn on the results of numerous computations carried out by my students. I now wish to express my appreciation to them all, particularly Susan McDougall Choy, Edward Croke, Gorman Gilbert, Jacques Ledent, Jorg Meise, and Joseph Prashker.

Several institutions have contributed to the development of this study over the past five years. I am indebted to the American–Yugoslav Project in Ljubljana for a very productive sabbatical leave in 1967 and for other support since that time. Grants from the US Economic Development Administration and the University of California's Committee on Research provided financial support for the study during 1968 and 1969. A grant from the National Science Foundation to Northwestern University's Urban Systems Engineering Center supported the study during its concluding two years in 1971 and 1972. Finally, my thanks go to the staff of the Center for Planning and Development Research (now the Institute of Urban and Regional Development) at the University of California, Berkeley, for their efforts to expedite the completion of early drafts of parts of this manuscript, several of which appeared as Working Papers, and to Connie Anderson for typing the final manuscript.

A. Rogers
Professor of Civil Engineering and Urban Affairs, Northwestern University

To Michael, Christopher, Kevin, and Laura

Contents

1	**Introduction**	1
1.1	Introduction	1
1.2	The statistical analysis of spatial dispersion	1
	1.2.1 Definitions	1
	1.2.2 The random spatial point pattern	2
1.3	Quadrat analysis	5
	1.3.1 The variance-mean ratio	6
	1.3.2 The chi-square goodness of fit test	7
	1.3.3 Some difficulties with the quadrat method	8
1.4	Nearest neighbor analysis	8
	1.4.1 The distribution of the nearest neighbor distance in a random point pattern	8
	1.4.2 Sample statistics	9
	1.4.3 Some difficulties with the nearest neighbor method	10
1.5	Organization of this study	11
2	**The fundamental component distributions**	12
2.1	Introduction	12
2.2	Random, regular, and clustered spatial dispersion	12
	2.2.1 Random spatial dispersion: the Poisson distribution	13
	2.2.2 Regular spatial dispersion: the binomial distribution	15
	2.2.3 Clustered spatial dispersion: the negative binomial distribution	16
2.3	Simulation of the fundamental component distributions	18
3	**Compound and generalized distributions**	21
3.1	Introduction	21
3.2	Definitions and notation	21
3.3	Compound Poisson distributions	23
	3.3.1 The Neyman Type A distribution	23
	3.3.2 The negative binomial distribution	24
3.4	Generalized Poisson distributions	25
	3.4.1 The Neyman Type A distribution	25
	3.4.2 The negative binomial distribution	25
3.5	Other compound and generalized distributions	26
	3.5.1 The Poisson-binomial distribution	27
	3.5.2 The Poisson-negative binomial distribution	27
3.6	Some properties of the derived compound and generalized distributions	28
4	**Parameter estimation**	31
4.1	Introduction	31
4.2	Moment and maximum likelihood estimation	31
	4.2.1 Moment estimators	31
	4.2.2 Maximum likelihood estimators	33

4.3	The Poisson distribution	33
	4.3.1 The moment estimator	33
	4.3.2 The maximum likelihood estimator	34
4.4	The binomial distribution	34
	4.4.1 The moment estimator	34
	4.4.2 The maximum likelihood estimator	34
4.5	The negative binomial distribution	35
	4.5.1 The moment estimators	35
	4.5.2 The maximum likelihood estimators	36
	4.5.3 Existence and efficiency of estimators	37
4.6	The Neyman Type A distribution	39
	4.6.1 The moment estimators	39
	4.6.2 The maximum likelihood estimators	39
	4.6.3 Existence and efficiency of estimators	41
4.7	The Poisson–binomial distribution	42
	4.7.1 The moment estimators	42
	4.7.2 The maximum likelihood estimators	43
	4.7.3 Existence and efficiency of estimators	45
4.8	The Poisson–negative binomial distribution	47
	4.8.1 The moment estimators	47
	4.8.2 The maximum likelihood estimators	49
	4.8.3 Existence and efficiency of estimators	50
4.9	Quadrat sampling and quadrat censusing	52
5	**Hypothesis testing: the chi-square goodness of fit test**	54
5.1	Introduction	54
5.2	The chi-square goodness of fit test	54
5.3	The power of the chi-square test	57
	5.3.1 The noncentral chi-square distribution	57
	5.3.2 The power function of the chi-square test	58
	5.3.3 The simulated power function	61
5.4	Some problems in the use of the chi-square test	67
	5.4.1 Estimation problems	68
	5.4.2 Grouping problems	68
6	**The structure of retail trade**	71
6.1	Introduction	71
6.2	Retailing	71
	6.2.1 Classes of retail goods	72
	6.2.2 Competition	73
	6.2.3 Retail location	74
6.3	The structure of retail trade: Ljubljana, Yugoslavia	75
	6.3.1 Productivity and size	78
	6.3.2 Size concentration	80
	6.3.3 Spatial concentration	85

7	**The spatial dispersion of retail trade in urban areas**	88
7.1	Introduction	88
7.2	Compound and generalized distributions as models of retail spatial dispersion	89
	7.2.1 The compound model	89
	7.2.2 The generalized model	89
7.3	The spatial dispersion of retail establishments in Ljubljana, Yugoslavia	90
	7.3.1 Intraurban comparisons	91
	7.3.2 Interurban comparisons	100
7.4	The problem of optimal quadrat size	105
7.5	The spatial dispersion of retail employment, space, and sales in Ljubljana, Yugoslavia	112
8	**Bivariate distributions**	116
8.1	Introduction	116
8.2	Random bivariate dispersion: the bivariate correlated Poisson distribution	116
8.3	Fundamental component distributions: a bivariate correlated negative binomial model	120
8.4	Bivariate compound distributions—another bivariate correlated negative binomial model	122
8.5	Bivariate generalized distributions—a bivariate correlated Neyman Type A model	124
8.6	Bivariate analysis of urban spatial dispersion in Ljubljana, Yugoslavia	126
9	**Spatial sampling**	132
9.1	Introduction	132
9.2	Probability sampling with marginal constraints: the Jessen method	133
	9.2.1 Definitions and notation	133
	9.2.2 The Jessen algorithm	136
9.3	Probability sampling with marginal constraints: quadrat spatial sampling	138
	9.3.1 A procedure for quadrat spatial sampling	139
	9.3.2 Quadrat spatial sampling of the retail and population distributions in Ljubljana, Yugoslavia	140
	9.3.3 Sensitivity tests of the quadrat spatial sampling procedure	142
9.4	Quadrat spatial sampling with estimated constraints	145
9.5	Conclusion	148
10	**Conclusion**	150
References		152
Author index		161
Subject index		162

Introduction

1.1 Introduction

In metropolitan regions across the world, efforts to cope with the pressing problems of growth and development have pointed to the need for a fuller understanding of the spatial structure of urban areas. Scholars in various disciplines have sought theories and methods which might contribute to an improved comprehension of human activities in space; their organization and distribution; their interlinkages; and the flows of people, goods, and information between them. Such endeavors have been supported by the belief that a better understanding of city systems will lead to a firmer foundation on which to base policy decisions designed to effect desired changes in these systems.

The geographical arrangement of activities in urban areas has received a great deal of attention from economists, geographers, city planners, and ecologists. Despite their efforts, however, we still do not have well developed and tested methods which allow us to express quantitatively the spatial patterns of various classes of urban activities, and to compare them on an intraurban and an interurban level. Recent efforts by quantitative geographers, and plant and animal ecologists, however, suggest that ultimately a body of techniques which provides such quantifications and comparisons may evolve out of a methodology that is commonly referred to as spatial point pattern analysis. A subclass of that methodology, quadrat analysis, is the concern of this book.

1.2 The statistical analysis of spatial dispersion

In recent years, location theorists have become increasingly aware of the need for quantitative descriptions of two-dimensional point patterns. Geographers in particular have recognized that pattern is the geometrical expression of location theory, and that techniques of pattern analysis therefore form an important methodological component of urban geography (Hudson and Fowler, 1966).

1.2.1 Definitions

The development of methods for describing and analyzing spatial patterns has been hampered by the lack of a precise definition of *pattern*, and how its properties differ from those of *shape* and *dispersion*. Every spatial arrangement of objects possesses these three properties and may be defined uniquely in terms of them. *Shape* is a two-dimensional characteristic of a spatial arrangement that is defined by a closed curve (Bunge, 1962) which delineates the collection of objects and provides an areal measure of their distribution. *Pattern* is a zero-dimensional characteristic of a spatial arrangement which describes the spacing of a set of objects with respect to one another (Hudson and Fowler, 1966). Finally *dispersion* may be viewed

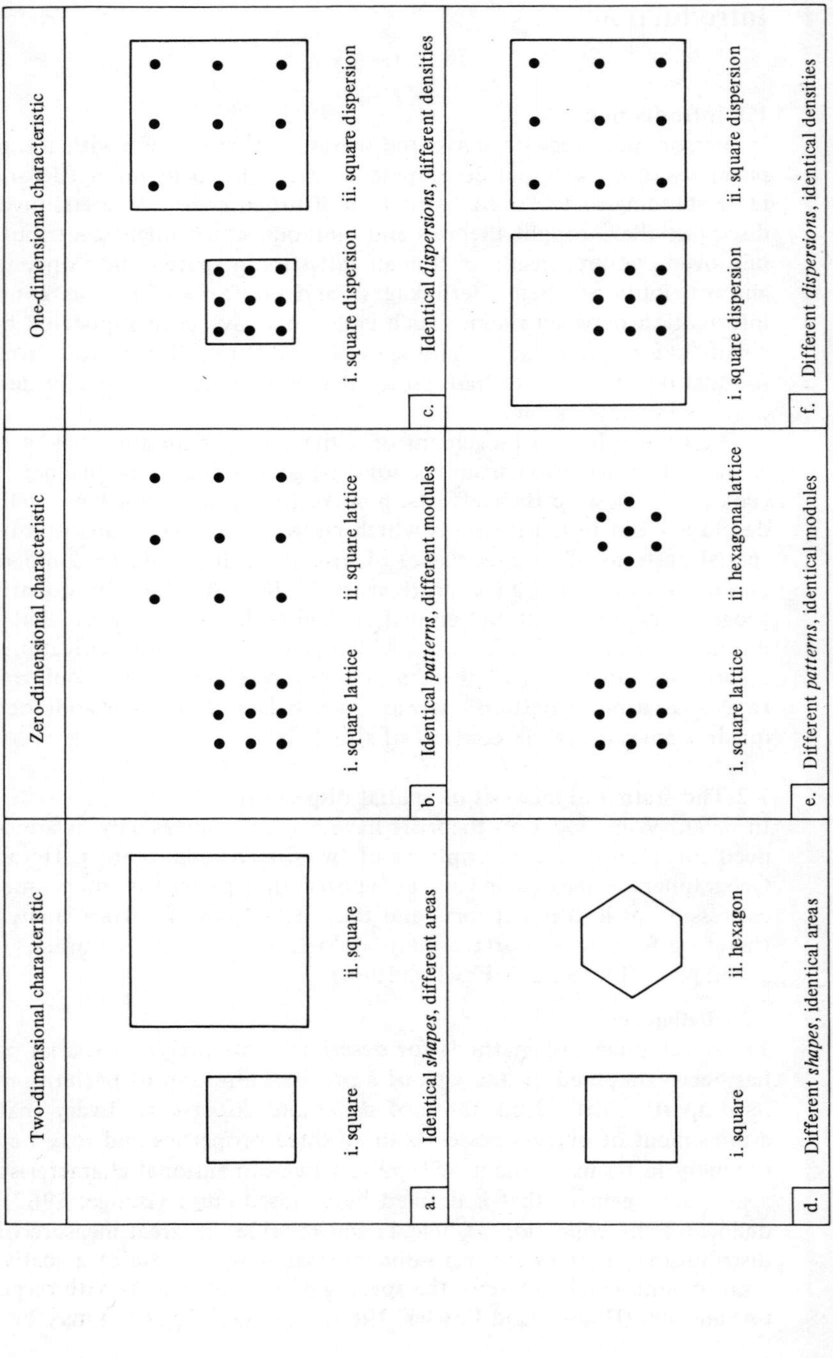

Figure 1.1. Characteristics of spatial arrangements.

as a one-dimensional characteristic of a spatial arrangement which measures the spacing of a set of objects in relation to one particular *shape* of a given area (McConnell, 1966). Thus we may view dispersion as an attribute of a *pattern* that is located within a particular *shape*, at a given density.

For a graphical illustration of the above distinctions between shape, pattern, and dispersion, let us turn to figure 1.1. Figure 1.1a provides an example of two identical shapes having different areas; figure 1.1b illustrates two identical patterns having different modules[1]; and figure 1.1c demonstrates that two identical patterns, placed into identical shapes, can lead to identical dispersions even when their densities are different. Figures 1.1d, 1.1e, and 1.1f, respectively, illustrate converse examples—namely, different shapes with identical areas, different patterns with identical modules, and different dispersions with identical shapes, patterns, and densities.

1.2.2 The random spatial point pattern

A fundamental concept which is used throughout this book is that of 'randomness'. It is necessary to define it unequivocally and in mathematical terms.

A random spatial point pattern is defined as the geographical disposition of points on a plane, generated by a realization of a spatial point process, that satisfies the following two conditions:

1 *Condition of equal probability.* Any point has an *equal probability* of occurring at any position on the plane. Therefore any subregion of the plane has the same probability of containing a point as that of any other subregion of equal area.
2 *Condition of independence.* The position of a point on the plane is *independent* of the position of any other point.

Any point located in accordance with the above two conditions is said to be located *randomly*, and the pattern generated by M such points is defined to be a *random spatial pattern* or, equivalently, a *realization* of a *random spatial point process*.

If M points are located randomly in a planar region, then the probability that a point falls within a particular subdivision of area A can be regarded as an event which occurs with probability λA, where λ is the density (number of points per unit area) of this random spatial point pattern. Consider, for example, a square subregion of area a in the plane and divide it into n very small square subdivisions. Assume that these square subdivisions are so small that the probability of more than one point occurring in them is insignificant and tends to zero as n becomes large. Then $A = a/n$, and the probability that a subdivision contains no points is $(1 - \lambda a/n)$. Since, from the n subdivisions, there are $\binom{n}{r}$ ways of

---
[1] By module we mean the average distance between a point and its nearest neighbor.

combining r subdivisions, each with only a single point, and because each of these combinations has the probability $(\lambda a/n)^r [(1 - \lambda a/n)]^{n-r}$ of occurring, the probability of finding r points in a square subregion of area a is given by the following expression:

$$P(R = r) = P(r) = \binom{n}{r} \left(\frac{\lambda a}{n}\right)^r \left(1 - \frac{\lambda a}{n}\right)^{n-r}$$

$$= \frac{n(n-1)\ldots(n-r+1)}{r!} \frac{(\lambda a)^r}{n^r} \left(1 - \frac{\lambda a}{n}\right)^n \left(1 - \frac{\lambda a}{n}\right)^{-r}$$

$$= \left(1 - \frac{1}{n}\right)\left(1 - \frac{2}{n}\right) \ldots \left(1 - \frac{r-1}{n}\right) \left(1 - \frac{\lambda a}{n}\right)^{-r}$$

$$\times \left[\left(1 - \frac{\lambda a}{n}\right)^n \frac{(\lambda a)^r}{r!}\right],$$

and as n becomes infinitely large, the terms outside the square brackets tend to unity and the terms inside the square brackets yield

$$P(r) = \exp(-\lambda a)\frac{(\lambda a)^r}{r!} \qquad (r = 0, 1, 2, \ldots,). \tag{1.1}$$

The probability distribution defined by equation (1.1) is known as the Poisson law with the single parameter λa. Its expected value is

$$E(r) = m_1 = \sum_{r=0}^{\infty} \exp(-\lambda a)\frac{(\lambda a)^r}{r!} = \lambda a \sum_{r=1}^{\infty} \exp(-\lambda a)\frac{(\lambda a)^{r-1}}{(r-1)!}$$

$$= \lambda a \sum_{x=0}^{\infty} \exp(-\lambda a)\frac{(\lambda a)^x}{x!}, \qquad \text{where } x = r - 1$$

$$= \lambda a. \tag{1.2}$$

Its second moment is given by

$$E(r^2) = \sum_{r=0}^{\infty} r^2 \exp(-\lambda a)\frac{(\lambda a)^r}{r!} = \sum_{r=0}^{\infty} r(r-1)\exp(-\lambda a)\frac{(\lambda a)^r}{r!}$$

$$+ \sum_{r=0}^{\infty} \exp(-\lambda a)\frac{(\lambda a)^r}{r!}$$

$$= \lambda^2 a^2 \sum_{r=2}^{\infty} \exp(-\lambda a)\frac{(\lambda a)^{r-2}}{(r-2)!} + \lambda a$$

$$= \lambda^2 a^2 + \lambda a.$$

Hence, the variance of the Poisson distribution is

$$\text{var}(r) = m_2 = E(r^2) - [E(r)]^2 = \lambda^2 a^2 + \lambda a - \lambda^2 a^2$$

$$= \lambda a. \tag{1.3}$$

Note that the mean and the variance of a Poisson distribution are equal.

Introduction

Figure 1.2b illustrates a random spatial point pattern which contains 52 points. In figure 1.2 we also present the two extreme point patterns that can be generated with the same 52 points. Figure 1.2a illustrates the case of maximum regularity, or *perfect regularity*, and figure 1.2c illustrates the case of maximum clustering, or *perfect clustering*. The *random* point pattern lies 'midway' between these two extreme distributions.

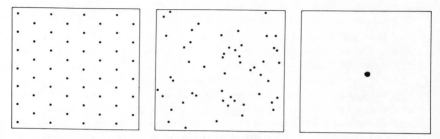

a. Perfectly regular point pattern   b. Random point pattern   c. Perfectly clustered point pattern

Figure 1.2. Perfectly regular, random, and perfectly clustered spatial point patterns ($M = 52$).

1.3 Quadrat analysis

The name quadrat analysis has been associated with a class of methods of spatial analysis which has been developed over the past thirty years, largely by plant and animal ecologists (Greig-Smith, 1964; Southwood, 1966). In the quadrat method, a planar study region is divided into a grid with cells of equal size, called *quadrats*, and the number of points in each cell, or in randomly selected cells, is noted. Quadrats are generally square shaped, and the analysis strives to obtain indications of the random or nonrandom character of the model that generated the spatial distribution being studied. An observed frequency distribution that does not conform with one expected from a random point process leads to a rejection of the hypothesis of randomness in favor of an alternative one of a *more regular*, or a *more clustered*, than random model, depending on the direction in which the observed values differ from those expected.

A regular point process would be expected to generate a large number of quadrats containing only a single point, some empty quadrats, and very few quadrats with more than one point in them. Conversely, a clustered point process would be expected to produce a very large number of empty quadrats, a few quadrats with one or two points, and several quadrats with many points in them. The random point process would be expected to produce results somewhere in between these two extremes.

Figure 1.3 illustrates the fit of a 10 by 10 grid of quadrats over the three point patterns in figure 1.2. In the case of the perfectly regular pattern, 52 out of the 100 quadrats have only a single point in them, 48 quadrats are empty, and none has more than a single point.

The perfectly clustered pattern, on the other hand, yields 99 empty quadrats and a single quadrat containing all the 52 points.

1.3.1 The variance-mean ratio

The variance of the Poisson distribution is equal to its mean. Hence the ratio of the variance to the mean for such distributions is equal to unity, and observed point dispersions may be measured for their departure from expected Poisson realizations by testing the significance of the difference between the observed variance-mean ratio and unity. This difference has a standard error of $[2/(N-1)]^{1/2}$, where N is the number of observations, and its significance may be statistically tested with a t-test with $N-1$ degrees of freedom (Greig-Smith, 1964). A realization yielding a variance-mean ratio greater than unity indicates a more clustered than random spatial point process, whereas one with a variance-mean ratio less than unity is more likely to result from a more regular than random model.

To illustrate the use of the variance-mean ratio test, we may compute this index for the three point dispersions in figure 1.3. The relevant statistics appear in table 1.1 [(2)].

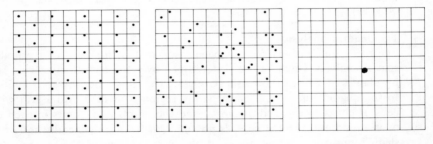

a. Perfectly regular   b. Random   c. Perfectly clustered

Figure 1.3. Quadrat coverage of perfectly regular, random, and perfectly clustered spatial point patterns ($M = 52$).

Table 1.1. Sample statistics for the variance-mean ratio test.

Perfectly regular	Random	Perfectly clustered
$\hat{m}_1 = 0\cdot 5200$	$\hat{m}_1 = 0\cdot 5200$	$\hat{m}_1 = 0\cdot 5200$
$\hat{m}_2 = 0\cdot 2521$	$\hat{m}_2 = 0\cdot 5148$	$\hat{m}_2 = 27\cdot 0400$
$\hat{m}_2/\hat{m}_1 = 0\cdot 4848$	$\hat{m}_2/\hat{m}_1 = 0\cdot 9899$	$\hat{m}_2/\hat{m}_1 = 52\cdot 0000$
$t^* = \dfrac{0\cdot 4848 - 1\cdot 0}{0\cdot 1421}$	$t^* = \dfrac{0\cdot 9899 - 1\cdot 0}{0\cdot 1421}$	$t^* = \dfrac{52\cdot 0 - 1\cdot 0}{0\cdot 1421}$
$= -3\cdot 6256$	$= -0\cdot 0711$	$= 358\cdot 9021$

[(2)] Henceforth we shall distinguish between theoretical and sample moments by denoting the former by m_i and the latter by u_i. Estimates of the theoretical moments will be further distinguished by a *caret* (or 'hat'), for example $\hat{m}_i$.

Each of the t statistics in table 1.1 is distributed as a t distribution with 99 degrees of freedom, and only the random dispersion is not significantly different from the realization that would be expected under a Poisson model, at the 5% level of significance.

1.3.2 The chi-square goodness of fit test

The chi-square goodness of fit test is an alternative method for measuring the departure of an observed spatial point dispersion from one which is random. This test is described in greater detail in chapter 5. Here we merely illustrate its application.

To apply the chi-square test, begin by estimating the population mean of the Poisson distribution, $m_1 = \lambda a$, with the sample mean, $\hat{m}_1 = u_1$. Then compute the expected frequencies:

$$NP(r) = N\exp(-\hat{m}_1)\frac{\hat{m}_1^r}{r!}, \qquad (r = 0, 1, 2, \ldots), \tag{1.4}$$

and measure the departure of observed from expected frequencies by means of the X^2 statistic:

$$X^2 = \sum_{r=0}^{W} \frac{[f_r - NP(r)]^2}{NP(r)}, \tag{1.5}$$

where $W+1$ denotes the number of frequency classes, $r = 0$ being one of them, f_r denotes the number of observations in the rth frequency class, N is the sample size $\left(\sum_{r=0}^{W} f_r = N\right)$, and $P(r)$ denotes the probability that an observation falls into the rth frequency class, under the hypothesis of random dispersion. Finally compare the computed X^2 statistic with tabulated percentiles for a chi-square distribution with $W-1$ degrees of freedom. Values of X^2 that exceed the critical value obtained from the chi-square tables lead to a rejection of the null hypothesis of randomness.

Table 1.2. Fitting the Poisson distribution to a perfectly regular, a random, and a perfectly clustered spatial point pattern ($M = 52$, $N = 100$).

Number of points per quadrat	Observed frequencies			Expected frequency with Poisson model ($\hat{\lambda}\hat{a} = 0\cdot5200$)
	perfectly regular	random	perfectly clustered	
0	48	59	99	59·45
1	52	32	0	30·92
2	0	7	0	8·04
3+	0	2	1	1·59
$N =$	100	100	100	100·00
$X^2 =$	–	0·08	64·96	
$P_{0\cdot05} =$	–	3·84	3·84	

Table 1.2 illustrates the application of the chi-square test to the three spatial point dispersions in figure 1.3. Note that in the case of the perfectly regular point pattern we have no degrees of freedom left to carry out the chi-square test, and observe that the perfectly clustered point pattern leads to an unqualified rejection of the null hypothesis of a random point process.

1.3.3 Some difficulties with the quadrat method

A host of theoretical and practical problems are associated with the quadrat method. Some of these will be considered in detail in this study. For example, the size of the quadrat influences the findings. This problem is discussed in chapter 7. Furthermore the method is density dependent. Hence it is used more appropriately as a technique for analyzing *dispersion* rather than pattern. The placement of the grid system over the point pattern can also influence the results.

Finally there are many vexing problems of inference associated with the fitting of nonrandom distributions. For example, several different probability models can give rise to the same expected frequency distribution—a point which will be illustrated in chapter 3. Moreover, a common rule of thumb in applications of the chi-square goodness of fit test is that observations be pooled in a way that provides an expectation of approximately 5 units for each frequency category. It will be seen in chapter 5 that this can lead to serious difficulties.

1.4 Nearest neighbor analysis

A method which gauges the departure from randomness of an observed spatial point pattern, but does not suffer from some of the faults of the quadrat method, is *nearest neighbor analysis*. This technique utilizes the distribution, in a random point pattern, of the distance between a point and its closest neighboring point. The derivation of this particular distribution, which appears below, is adapted from the argument in Clark and Evans (1954a).

1.4.1 The distribution of the nearest neighbor distance in a random point pattern

Consider the distance of any point to its nearest neighbor in a random pattern containing M points. First, center a circular quadrat of area a over the point. The probability of finding r points in this quadrat is given by the Poisson distribution

$$P(r) = \exp(-\lambda a) \frac{(\lambda a)^r}{r!} .$$

Hence, the probability of finding no points is

$$P(0) = \exp(-\lambda a) ,$$

or, if the radius of the circular quadrat is d, then

$$P(0) = \exp(-\lambda \pi d^2) . \tag{1.6}$$

But this is also the probability that the distance D to the nearest neighboring point is greater than d. Thus

$$P(D > d) = P(0) = \exp(-\lambda \pi d^2).$$

Consequently,

$$P(D \leq d) = 1 - P(0) = 1 - \exp(-\lambda \pi d^2) = F(d), \qquad (1.7)$$

where $F(d)$ is the cumulative distribution function of the nearest neighbor distance, d.

Differentiating equation (1.7), we obtain the associated probability density function

$$f(d) = F'(d) = 2\lambda \pi d \exp(-\lambda \pi d^2), \qquad (1.8)$$

and conclude, therefore, that

$$E(d) = \frac{1}{2\lambda^{1/2}}, \qquad (1.9)$$

and

$$\mathrm{var}(d) = \frac{4-\pi}{4\lambda\pi}. \qquad (1.10)$$

1.4.2 Sample statistics

Since, in a random point pattern, each nearest neighbor distance is distributed independently from all others and follows the distribution defined by equation (1.8), we may conclude that the average nearest neighbor distance, given by

$$\bar{d} = \frac{1}{M} \sum_{i=1}^{M} d_i, \qquad (1.11)$$

is distributed approximately as the normal distribution, with mean

$$E(\bar{d}) = \frac{1}{2\lambda^{1/2}} \quad \text{and variance} \quad \mathrm{var}(\bar{d}) = \frac{4-\pi}{4\lambda\pi M}.$$

Thus

$$\phi = \frac{\bar{d} - E(\bar{d})}{[\mathrm{var}(\bar{d})]^{1/2}} \qquad (1.12)$$

approximately follows a normal distribution with zero mean and unit variance.

Clark and Evans (1954a) suggest the use of the index

$$D^* = \frac{\bar{d}}{E(\bar{d})} \qquad (1.13)$$

to compare different point patterns. They show that a perfectly regular point pattern leads to a $D^* = 2 \cdot 1491$, a random point pattern to a $D^* = 1$, and, of course, a perfectly clustered pattern to a $D^* = 0$.

To illustrate the application of the nearest neighbor test, this index may be computed for the three point patterns in figure 1.2, which yields an estimate of λ that is equal to $0 \cdot 0052$, and the following sample statistics[3]:

	$\bar{d}$	D^*	ϕ
Perfectly regular	15·00	2·16	$\dfrac{15 \cdot 00 - 6 \cdot 93}{0 \cdot 5026} = 16 \cdot 06$
Random	7·59	1·09	$\dfrac{7 \cdot 59 - 6 \cdot 93}{0 \cdot 5026} = 1 \cdot 31$
Perfectly clustered	0·00	0·00	$\dfrac{0 \cdot 00 - 6 \cdot 93}{0 \cdot 5026} = -13 \cdot 79$.

Once again, only the random dispersion is not significantly different from the realization that would be expected under a Poisson model, at the 5% level of significance.

1.4.3 Some difficulties with the nearest neighbor method

Although free of some of the difficulties which were associated with the quadrat method (such as the problem of quadrat size), nearest neighbor analysis does suffer from several serious problems. Some problems of quadrat methods, such as density dependence, are shared by the nearest neighbor technique. Others are unique to the method itself. For example, sole reliance on the nearest neighbor distance may lead to erroneous conclusions if the point pattern has a configuration of 'clumps', as in figure 1.4, which yields a nearest neighbor distance of zero, thereby indicating perfect clustering when it does not exist. To deal with this problem, Thompson (1956) and others have suggested the use of subareas or nth nearest neighbors. Thus the average distances to the first, second, third, fourth, and fifth nearest neighbors in the spatial point patterns in figure 1.2 are shown in table 1.3.

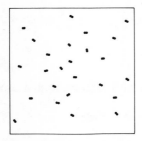

Figure 1.4. Pattern of two-point clumps ($M = 52$).

---

[3] In computing the nearest neighbor statistics we measured distances between points to the nearest tenth of a quadrat edge. Hence the estimate of λ used here is $0 \cdot 0052$, or $1/10^2$ of the corresponding estimate of λ used in subsection 1.3.2.

Introduction

Perhaps the most serious limitation of nearest neighbor analysis is the difficulty of obtaining the probability distributions of nearest neighbor distances in spatial point processes other than the random one. Thus, although nonrandom patterns may be ranked according to their degree of regularity or clustering, it is exceedingly difficult to infer much about the probability models that generated them. This is the principal reason for our adoption of the quadrat method as the technique for analyzing spatial dispersion in this study.

Table 1.3. Average distances to the first five nearest neighbors in the perfectly regular, random, and perfectly clustered point patterns of figure 1.2.

	Perfectly regular[a]	Random	Perfectly clustered
First:	15·00	7·59	0·00
Second:	15·00	11·49	0·00
Third:	15·43	14·16	0·00
Fourth:	17·12	16·57	0·00
Fifth:	19·17	18·87	0·00

[a] Although, in theory, the first six nearest neighbor distances should be equal in the perfectly regular point pattern, the existence of a 'boundary effect' in figure 1.2 is evident in the third, fourth, and fifth nearest neighbor distances. The same effect occurs in the higher nearest neighbor distances of the random point pattern.

1.5 Organization of this study

The growing interest in the quantitative description and analysis of spatial pattern has generated a considerable literature in several disciplines. The current state of the art, however, still does not allow us to measure adequately the fundamental zero-dimensional attribute of point patterns—namely the geometrical arrangement of a set of points which remains invariant under transformations of shape and dispersion. Nevertheless, techniques falling under the general category of quadrat analysis can be satisfactorily applied to measure the spatial *dispersion* of a set of points within a predefined study area. This study provides a summary of such techniques and illustrates their application to an important problem in urban analysis—the description of the spatial dispersion of retail establishments in urban areas.

The organizational structure of this monograph divides the subject matter into four parts. Part 1 focuses on probability models in quadrat analysis. Part 2 considers the problem of statistical inference. Part 3 applies quadrat analysis to the study of retail spatial structure, and part 4 deals with certain special topics.

The fundamental component distributions

2.1 Introduction
In order to determine whether the dispersion of a particular set of points relative to a predefined study region is regular, random, or clustered, we must first define in mathematical terms the distributional form such dispersions are likely to exhibit. The usual procedure is to adopt the Poisson distribution as the mathematical definition of 'randomness'. Following Arbous and Kerrich (1951), we shall now rederive this distribution in a way that will allow us to extend our analysis to nonrandom dispersions.

2.2 Random, regular, and clustered spatial dispersion
Imagine a study region that has been gridded into square cells of a given dimension. Assume that initially (that is, at time $t = 0$), none of the cells contains any points, and let $p(r, t)$ be the probability that an individual grid cell has r points by time t. Assume that during the time interval $(t, t+dt)$ a point locates in a particular cell, which already has r points, with probability $f(r, t)dt$, and that the time interval is short enough for no more than one point to locate in a given cell during a single time interval. It then follows that:

$$p(0, t+dt) = p(0, t)[1 - f(0, t)dt],$$

$$p(r, t+dt) = p(r, t)[1 - f(r, t)dt] + p(r-1, t)f(r-1, t)dt \qquad (r = 1, 2, 3, ...),$$

and, subtracting $p(r, t)$ from both sides of the above equations, dividing by dt, and taking the limit as $dt \to 0$, we have:

$$\frac{\partial}{\partial t}p(0, t) = -f(0, t)p(0, t),$$

$$\frac{\partial}{\partial t}p(r, t) = -f(r, t)p(r, t) + f(r-1, t)p(r-1, t) \qquad (r = 1, 2, 3, ...). \qquad (2.1)$$

Let us multiply the first equation in (2.1) by s^0, the second by s, the third by s^2, and, in general, the nth by s^{n-1}. Summing the resulting equations, we obtain

$$\frac{\partial}{\partial t}\left[\sum_{r=0}^{\infty} p(r, t)s^r\right] = (s-1)\left[\sum_{r=0}^{\infty} f(r, t)p(r, t)s^r\right],$$

or, more compactly,

$$\frac{\partial}{\partial t}G(s; t) = (s-1)L(s; t), \qquad (2.2)$$

where

$$G(s;t) = \sum_{r=0}^{\infty} p(r, t)s^r$$

is the probability generating function of the random variable r,[4] and

$$L(s;t) = \sum_{r=0}^{\infty} f(r, t)p(r, t)s^r \ .$$

To find $G(s;t)$ we need to solve the single differential equation in equation (2.2). Depending on the assumption we make concerning $f(r, t)$, our resulting distribution will be either a *random* one, a *regular* one, or a *clustered* one. Note that $f(r, t)$ is a probability that is contingent on the value of r. It expresses the probability that a cell with r points will obtain yet another one in the time interval $(t, t+dt)$. If this probability is independent of the number of points already in the cell, we will refer to the spatial dispersion as *random* dispersion. If, on the other hand, it declines as the number of existing points in the cell increases, then we will define the spatial dispersion to be a *regular* one. Finally, if the probability increases as the number of existing points in the cell increases, we will refer to the resulting spatial dispersion as *clustered* dispersion. In the following sections we shall be more specific and assume functional values for $f(r, t)$, and derive the distributional consequences of these assumptions. The resulting distributions will be called *the fundamental component distributions of quadrat analysis*, because most of the other more complex distributions that are commonly used in quadrat analysis can be derived as particular mixtures or combinations of these basic distributions.

2.2.1 Random spatial dispersion: the Poisson distribution

Assume that the probability that a cell receives a point during the time interval $(t, t+dt)$ is completely independent of the number of such points already in the cell. Then

$$f(r, t) = f(t) \ , \tag{2.3}$$

$$L(s;t) = \sum_{r=0}^{\infty} f(t)p(r, t)s^r = f(t)G(s;t) \ ,$$

and equation (2.2) becomes

$$\frac{\partial}{\partial t}G(s;t) = (s-1)f(t)G(s;t) \ ,$$

---

[4] For a full discussion of the role and use of probability generating functions in discrete probability theory see Feller (1957, pp.248–261).

with the solution
$$G(s;t) = \exp\left[(s-1)\int_0^t f(t')dt'\right].$$
Hence for any point in time, $\bar{t}$ say,
$$G(s;\bar{t}) = G(s) = \exp[\lambda a(s-1)], \qquad (2.4)$$
where
$$\lambda a = \int_0^{\bar{t}} f(t')dt'.$$

Equation (2.4) is easily recognizable as the probability generating function (p.g.f.) of the Poisson distribution with parameter λa. Hence
$$p(r,\bar{t}) = P(r) = \exp(-\lambda a)\frac{(\lambda a)^r}{r!} \qquad (r = 0, 1, 2, ...). \qquad (2.5)$$

As a check let us compute the p.g.f. of the above Poisson distribution. By definition,
$$G(s) = \sum_{r=0}^{\infty} P(r)s^r.$$
Hence
$$G(s) = \sum_{r=0}^{\infty} \exp(-\lambda a)\frac{(\lambda a)^r}{r!}s^r = \exp(-\lambda a)\sum_{r=0}^{\infty}\frac{(\lambda a s)^r}{r!}$$
$$= \exp(-\lambda a)\exp(\lambda a s)$$
$$= \exp[\lambda a(s-1)],$$
using the standard relationship
$$\exp x = \sum_{n=0}^{\infty} \frac{x^n}{n!}.$$

Using the well-known relationships
$$E(r) = m_1 = \frac{d}{ds}G(s)\bigg|_{s=1} = G'(1),$$
and
$$\mathrm{var}(r) = m_2 = G''(1) + G'(1) - [G'(1)]^2,$$
where the double prime denotes the second derivative with respect to s, we may derive the mean and variance of the Poisson distribution:
$$m_1 = \lambda a$$
$$m_2 = \lambda a.$$

Note once again that the variance–mean ratio of this distribution is always unity, that is, $m_2/m_1 = \lambda a/\lambda a = 1$.

2.2.2 Regular spatial dispersion: the binomial distribution

For the next derivation, assume that the probability that a point locates in a cell is independent of time and decreases linearly with the number of points already in that cell. More specifically, let c/b be an integer and

$$f(r, t) = c - br \quad \text{for } c > br \geq 0$$
$$= 0 \quad \text{otherwise.} \tag{2.6}$$

Then

$$L(s;t) = \sum_{r=0}^{\infty}(c-br)p(r,t)s^r = c\sum_{r=0}^{\infty}p(r,t)s^r - b\sum_{r=0}^{\infty}rp(r,t)s^r$$

$$= cG(s;t) - bs\frac{\partial}{\partial s}G(s;t),$$

and equation (2.2) becomes

$$\frac{\partial}{\partial t}G(s;t) = (s-1)\left[cG(s;t) - bs\frac{\partial}{\partial s}G(s;t)\right], \tag{2.7}$$

with the solution

$$G(s;t) = \{\exp(-bt) - [\exp(-bt) - 1]s\}^{c/b}.$$

Hence for any point in time $\bar{t}$ say, we may make the following substitutions:

$$p = 1 - \exp(-b\bar{t}) \quad \text{and} \quad n = c/b,$$

to find

$$G(s;\bar{t}) = G(s) = (1 - p + ps)^n. \tag{2.8}$$

Equation (2.8) is the probability generating function of the binomial distribution

$$P(r) = \binom{n}{r}p^r(1-p)^{n-r} \quad (r = 0, 1, 2, ..., n). \tag{2.9}$$

As a check we may compute the p.g.f. of the above binomial distribution as follows:

$$G(s) = \sum_{r=0}^{\infty}P(r)s^r = \sum_{r=0}^{n}\binom{n}{r}p^r(1-p)^{n-r}s^r$$

$$= \sum_{r=0}^{n}\binom{n}{r}(ps)^r(1-p)^{n-r}$$

$$= (1 - p + ps)^n.$$

Differentiating to derive the mean and variance, we obtain

$$E(r) = m_1 = G'(1) = np,$$

and

$$\text{var}(r) = m_2 = G''(1) + G'(1) - [G'(1)]^2 = np(1-p).$$

Observe that the variance–mean ratio of this distribution is $1-p$, which is always less than unity.

When n is large and p is small, that is, when $n \to \infty$ and $p \to 0$, such that $np = \lambda a$, the Poisson distribution provides a reasonably close approximation to the binomial distribution. This may be proved by means of a limiting argument on the probability generating function of the binomial distribution. For the binomial distribution,

$$G(s) = (1-p+ps)^n .$$

If we let $n \to \infty$ and $p \to 0$, such that $np = \lambda a$ remains fixed, then

$$(1-p+ps)^n = \left[1 - \frac{\lambda a(1-s)}{n}\right]^n ,$$

and

$$\lim_{\substack{n \to \infty \\ p \to 0}} \left[1 - \frac{\lambda a(1-s)}{n}\right]^n = \exp[\lambda a(s-1)] ,$$

which has already been shown to be the probability generating function of the Poisson distribution.

2.2.3 Clustered spatial dispersion: the negative binomial distribution

For the final derivation, assume that the probability that a point locates in a cell is independent of time and increases linearly with the number of points already in that cell. In particular, assume that

$$f(r, t) = c + br \qquad (c > 0, b > 0) . \tag{2.10}$$

Then

$$L(s;t) = \sum_{r=0}^{\infty} (c+br)p(r,t)s^r = c \sum_{r=0}^{\infty} p(r,t)s^r + b \sum_{r=0}^{\infty} rp(r,t)s^r$$

$$= cG(s;t) + bs\frac{\partial}{\partial s}G(s;t) ,$$

and equation (2.2) becomes

$$\frac{\partial}{\partial t}G(s;t) = (s-1)\left[cG(s;t) + bs\frac{\partial}{\partial s}G(s;t)\right] , \tag{2.11}$$

with the solution

$$G(s;t) = [\exp bt - (\exp bt - 1)s]^{-c/b} .$$

For any point in time, $\bar{t}$ say, we may make the substitutions

$$p = \exp b\bar{t} - 1 \quad \text{and} \quad k = c/b ,$$

to find

$$G(s;\bar{t}) = G(s) = (1+p-ps)^{-k} . \tag{2.12}$$

Equation (2.12) is the probability generating function of the negative binomial distribution

$$P(r) = \binom{k+r-1}{r} \left(\frac{p}{1+p}\right)^r \left(\frac{1}{1+p}\right)^k . \qquad (2.13)$$

We may compute the p.g.f. of the above distribution to find

$$G(s) = \sum_{r=0}^{\infty} P(r)s^r = \sum_{r=0}^{\infty} \binom{k+r-1}{r} \left(\frac{p}{1+p}\right)^r \left(\frac{1}{1+p}\right)^k s^r$$

$$= \left(\frac{1}{1+p}\right)^k \sum_{r=0}^{\infty} \binom{k+r-1}{r} \left[\left(\frac{p}{1+p}\right)s\right]^r$$

$$= \left(\frac{1}{1+p}\right)^k \left[1 - \left(\frac{p}{1+p}\right)s\right]^{-k}$$

$$= (1+p-ps)^{-k} .$$

Differentiating $G(s)$ to derive the mean and variance, we find

$$E(r) = m_1 = G'(1) = kp ,$$

and

$$\text{var}(r) = m_2 = G''(1) + G'(1) - [G'(1)]^2 = kp(1+p) .$$

Note that the variance-mean ratio for this distribution is $1+p$, which is always greater than unity.

When k is large and p is small, that is, when $k \to \infty$ and $p \to 0$, such that $kp = \lambda a$, the negative binomial distribution approaches the Poisson distribution. As with the binomial case, this can be established by an argument using probability generating functions. For the negative binomial distribution,

$$G(s) = (1+p-ps)^{-k} .$$

Hence, if we let $k \to \infty$ and $p \to 0$, such that $kp = \lambda a$ remains fixed, then

$$(1+p-ps)^{-k} = \left[1 + \frac{\lambda a(1-s)}{k}\right]^{-k} ,$$

and

$$\lim_{\substack{k \to \infty \\ p \to 0}} \left[1 + \frac{\lambda a(1-s)}{k}\right]^{-k} = \frac{1}{\exp \lambda a(1-s)}$$

$$= \exp \lambda a(s-1) ,$$

which is the p.g.f. of the Poisson distribution.

Occasionally it is more convenient to operate with the following alternative forms of the negative binomial distribution:

$$P(r) = \binom{k+r-1}{r}\left(\frac{w}{w+k}\right)^r\left(\frac{k}{w+k}\right)^k \quad (w = kp), \tag{2.14}$$

$$= \binom{k+r-1}{r} p^r q^{-k-r} \quad (q = 1+p), \tag{2.15}$$

$$= \binom{k+r-1}{r} Q^r P^k \quad (Q = 1-P); \tag{2.16}$$

and their respective probability generating functions:

$$G(s) = \left(1 + \frac{w}{k} - \frac{ws}{k}\right)^{-k}, \tag{2.17}$$

$$= (1+p-ps)^{-k} = (q-ps)^{-k}, \tag{2.18}$$

$$= \left(\frac{P}{1-Qs}\right)^k. \tag{2.19}$$

2.3 Simulation of the fundamental component distributions[5]

It is frequently desirable to obtain theoretical spatial point patterns that have specific distributional properties. Such simulated patterns may be used, for example, to test the influence of alternative quadrat sizes. Moreover they provide the synthetically created data that are so useful in the study of the efficiency of alternative parameter estimation methods. The above derivations of the Poisson, binomial, and negative binomial distributions suggest a way in which such component distributions may be simulated.

Recall the following three fundamental stochastic processes defined by equations (2.3), (2.6), and (2.10), respectively:

Poisson: $f(r, t) = f(t) = c$, say, $\quad c > 0$;

Binomial: $f(r, t) = c - br \quad$ for $c > br \geq 0$,
$\qquad\qquad\quad = 0 \qquad\qquad$ otherwise;

Negative binomial: $f(r, t) = c + br$, $\quad$ for $b > 0$ and $c > 0$.

If M points are located in a grid according to one of these stochastic processes, the resulting point dispersion should approximate a realization of the process selected. Thus, for example, if in a grid of N squares M points are located according to the process defined by equation (2.3), the resulting point dispersion should be an approximate realization of the Poisson distribution with mean equal to M/N. Alternatively, if each of the

---

[5] For a more detailed exposition of computer generated spatial patterns, see Martin et al. (1969).

The fundamental component distributions

Table 2.1. Observed and expected distributions of the simulated patterns.
B: binomial distribution; P: Poisson distribution; NB: negative binomial distribution.

No. of points per cell	Perfectly regular[a]			Regular[a]			Random[a]			Clustered				Perfectly clustered			
	Observed frequency	B $\hat{n}=1$ $\hat{p}=0.5200$	P $\hat{\lambda}=0.5200$	Observed frequency	B $\hat{n}=2$ $\hat{p}=0.2600$	P $\hat{\lambda}=0.5200$	Observed frequency	B $\hat{n}=3$ $\hat{p}=0.1733$	P $\hat{\lambda}=0.5200$	Observed frequency	B $\hat{n}=5$ $\hat{p}=0.1040$	P $\hat{\lambda}=0.5200$	NB $\hat{w}=k\hat{p}=0.5200$ $\hat{k}=0.6781$	Observed frequency	B $\hat{n}=52$ $\hat{p}=0.0100$	P $\hat{\lambda}=0.5200$	NB $\hat{w}=k\hat{p}=0.5200$ $\hat{k}=0.0102$
0	48	48.00	59.45	54	54.76	59.45	59	56.49	59.45	67	57.75	59.45	67.98	99	59.30	59.45	96.05
1	52	52.00	40.55	40	38.48	30.92	32	35.54	30.92	23	33.51	30.92	20.01	0	31.15	30.92	0.96
2				6	6.76	9.63	7	7.45⎫	8.04⎫	5	7.78	8.04	7.29	0	8.02⎫	8.04⎫	0.48⎫
3							2	0.52⎭	1.59⎭	2	0.90	1.39	2.82	0	1.35⎟	1.39⎟	0.31⎟
4										2	0.05	0.18⎫	1.13⎫	0	0.17⎟	0.18⎟	0.23⎟
5										1		0.02⎭	0.78⎭	0	0.02⎭	0.02⎭	0.18⎟
6														0			0.15⎟
7														0			0.13⎟
8														0			0.11⎟
9														0			0.09⎭
10+														1			1.31
X^2	–	0.16	–		0.60	4.54		4.96	3.00		[2.05][b]		65.39	64.96	[2.75][b]		
$P_{0.05}$	–	3.84	–		3.84	3.84		3.84	3.84		5.99		3.84	3.84	3.84		

$\hat{m}_1 = 0.5200$		$\hat{m}_1 = 0.5200$	$\hat{m}_1 = 0.5200$	$\hat{m}_1 = 0.5200$	$\hat{m}_1 = 0.5200$
$\hat{m}_2 = 0.2521$	$D^* = 2.16$	$\hat{m}_2 = 0.3733$ $D^* = 1.44$	$\hat{m}_2 = 0.5148$ $D^* = 1.09$	$\hat{m}_2 = 0.9188$ $D^* = 0.75$	$\hat{m}_2 = 27.0400$ $D^* = 0.00$
$\hat{m}_2/\hat{m}_1 = 0.4848$		$\hat{m}_2/\hat{m}_1 = 0.7180$	$\hat{m}_2/\hat{m}_1 = 0.9899$	$\hat{m}_2/\hat{m}_1 = 1.7669$	$\hat{m}_2/\hat{m}_1 = 52.0000$

[a] Moment estimators do not exist for the negative binomial distribution because the sample variance is smaller than the sample mean.
[b] $[X^2] = X^2$ statistic computed with grouping $\geqslant 1$ instead of $\geqslant 5$.

M points is located, sequentially, according to equation (2.10), the resulting point dispersion should be an approximate realization of the negative binomial distribution with parameters that are functions of the constants c and b.

The above procedures yield point dispersions, in other words the number of points in each quadrat. However, they do not provide point patterns. To obtain the Poisson, binomial, and negative binomial patterns illustrated in figures 1.2 and 2.1, therefore, we proceeded as follows. First, we located 52 points into 100 quadrats according to the rules specified by equations (2.3), (2.6), and (2.10), respectively. Second, within each quadrat we located the requisite number of points randomly. The resulting realizations seem to be adequate representatives of their respective asymptotic distributions, according to table 2.1.

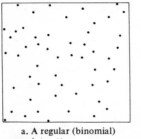   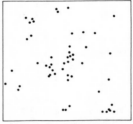

a. A regular (binomial) point pattern

b. A clustered (negative binomial) point pattern

Figure 2.1. A regular (binomial) and a clustered (negative binomial) spatial point pattern ($M = 52$).

3

Compound and generalized distributions

3.1 Introduction
In the preceding chapter, spatial dispersions which exhibited a quadrat sample variance that exceeded the quadrat sample mean were said to be *clustered*. In this chapter, we examine further this class of spatial dispersions and, in the process, show that its most important theoretical distributions may be derived using distinctly different sets of assumptions. We shall see later that this poses serious problems because it means that one cannot rely solely on empirical point dispersions in making inferences regarding the stochastic process that governs their geographical arrangement.

Two alternative stochastic processes that produce clustered spatial dispersion will be considered: *compound* processes and *generalized* processes. In the former, the clustered dispersion arises out of a basic inhomogeneity in the population being studied. In the latter, the clustering is a consequence of a fundamental locational affinity that members of a particular population have for each other.

3.2 Definitions and notation
Following Gurland (1957), we shall adopt the following definitions for compound and generalized distributions:

Compound distribution. If R_1 is a random variable with probability density function $P_1(r_1|R_2)$, for a given value of R_2, and if R_2 is regarded as a random variable with probability density function $P_2(r_2)$, then the random variable $R \equiv R_1 \wedge R_2$, with probability density function

$$P(r) = P(R = r) = \int_{-\infty}^{\infty} P_1(r|cr_2)P_2(r_2)\,dr_2 \,, \tag{3.1}$$

is called the *compound R_1 distribution* with respect to the compounder R_2. Here c is an arbitrary constant which is introduced merely to simplify our subsequent manipulations of the derived distributions. The constant is always consistent with the requirements of the distributions involved.

Generalized distribution. If R_1 is a random variable with probability density function $P_1(r_1)$, and if R_2 is a random variable with probability density function $P_2(r_2)$, then the random variable

$$R \equiv R_1 \vee R_2 = \sum_{i=1}^{R_1} R_{2i} = R_{21} + R_{22} + \ldots + R_{2R_1} \,, \tag{3.2}$$

with probability density function

$$P(r) = P(R = r) = \sum_{r_1 = 0}^{\infty} P_{2*}(r|r_1)P_1(r_1) \,, \tag{3.3}$$

where $P_{2*}(r|r_1)$ denotes the r_1-fold convolution of r_2 with itself[6], and $P(r)$ is called the *generalized R_1 distribution* with respect to the generalizer R_2, in which $R_{2i}(i = 1, 2, ..., R_1)$ is a family of *independent and identically distributed* integer-valued random variables that are independent of R_1.

The above definitions of compound and generalized distributions may be verbalized as follows. A compound distribution arises out of the presence of heterogeneity in a parameter of a particular probability distribution that describes a given population. If the values of the parameter vary in a specified way, then the resultant 'mixing' of the probability distribution that is involved leads to an 'apparent', but not 'real', spatial clustering. Real clustering does occur in a generalized distribution, however, for here the population is assumed to be spatially arranged in the form of clusters of varying size.

For the compound distribution in equation (3.1) we can establish the following useful relationship involving probability generating functions. Let $G(s)$ denote the p.g.f. of the random variable R, and let $P(r)$ denote its probability density function. Then, by definition,

$$G(s) = \sum_{r=0}^{\infty} P(r)s^r, \tag{3.4}$$

and the random variable $R \equiv R_1 \wedge R_2$, with the probability density function in equation (3.1), has the p.g.f.

$$\begin{aligned} G(s) &= \sum_{r=0}^{\infty} P(r)s^r = \sum_{r=0}^{\infty} \left[\int_{-\infty}^{\infty} P_1(r|cr_2) P_2(r_2) \mathrm{d}r_2 \right] s^r \\ &= \int_{-\infty}^{\infty} \left[\sum_{r=0}^{\infty} P_1(r|cr_2) s^r \right] P_2(r_2) \mathrm{d}r_2 \\ &= \int_{-\infty}^{\infty} G_1(s|cr_2) P_2(r_2) \mathrm{d}r_2 \,. \end{aligned} \tag{3.5}$$

Similarly, for the generalized distribution in equation (3.3) we can establish the following useful relationship involving probability generating functions:

$$\begin{aligned} G(s) &= \sum_{r=0}^{\infty} P(r)s^r = \sum_{r=0}^{\infty} \left[\sum_{r_1=0}^{\infty} P_{2*}(r|r_1) P_1(r_1) \right] s^r \\ &= \sum_{r_1=0}^{\infty} P_1(r_1) \sum_{r=0}^{\infty} P_{2*}(r|r_1) s^r \\ &= \sum_{r_1=0}^{\infty} P_1(r_1) G_{2*}(s|r_1) \,, \end{aligned} \tag{3.6}$$

where $G_{2*}(s|r_1)$ is the p.g.f. of $P_{2*}(r|r_1)$.

---

[6] See Feller (1957), pp.250-253.

For fixed r_1, the distribution of R is the r_1-fold convolution of $P_2(r_2)$ with itself. Hence $G_{2*}(s|r_1) = [G_2(s)]^{r_1}$. Consequently

$$G(s) = \sum_{r_1=0}^{\infty} P_1(r_1) G_{2*}(s|r_1) = \sum_{r_1=0}^{\infty} P_1(r_1)[G_2(s)]^{r_1}$$
$$= G_1[G_2(s)] . \tag{3.7}$$

3.3 Compound Poisson distributions

Let R_1 be a Poisson distributed random variable with mean λa. Then the random variable $R \equiv R_1 \wedge R_2$, with the probability density function in equation (3.1), is called a compound Poisson variable, and its distribution is referred to as a *compound Poisson distribution*. Whence

$$G_1(s|cr_2) = G_1(s|a\lambda) = \exp[\lambda a(s-1)] ,$$

where c and r_2 have been set equal to a and λ, respectively, and

$$G(s) = \int_{-\infty}^{\infty} G_1(s|cr_2) P_2(r_2) dr_2$$
$$= \int_{-\infty}^{\infty} \exp[\lambda a(s-1)] P_2(\lambda) d\lambda$$
$$= M_2[a(s-1)] , \tag{3.8}$$

where $M_2(\theta)$ is the moment generating function (m.g.f.) of the probability density function $P_2(r_2)$[7].

3.3.1 The Neyman Type A distribution

One of the simplest and most common compound Poisson distributions is obtained by introducing another Poisson variable as the compounder, R_2. Then

$$P_2(r_2) = \exp(-v)\frac{v^{r_2}}{r_2!} ,$$

and

$$M_2(\theta) = \int_{-\infty}^{\infty} \exp(r_2\theta) P_2(r_2) dr_2 = \sum_{r_2=0}^{\infty} \exp(r_2\theta)\exp(-v)\frac{v^{r_2}}{r_2!}$$
$$= \exp[v(\exp\theta - 1)] . \tag{3.9}$$

Whence

$$G(s) = M_2[a(s-1)] = \exp\langle v\{\exp[a(s-1)] - 1\}\rangle , \tag{3.10}$$

[7] Moment generating functions play a role in continuous probability theory that is analogous to the role played by probability generating functions in discrete probability theory. However, whereas the former can be used with both classes of probability distributions, the latter are usually applied only to discrete probability distributions. [See Parzen (1960), pp.215-223.]

which is the p.g.f. of the *Neyman Type A distribution* (Neyman, 1939) with density function

$$P(r) = \exp(-v)\frac{a^r}{r!} \sum_{i=0}^{\infty} \frac{i^r}{i!} [v\exp(-a)]^i \qquad (0^0 \equiv 1),\qquad(3.11)$$

and mean and variance:

$$E(r) = m_1 = va,$$

$$\text{var}(r) = m_2 = va(a+1).$$

3.3.2 The negative binomial distribution

The negative binomial distribution may be derived as a compound Poisson distribution that is compounded by the gamma distribution

$$P_2(r_2) = P_2(\lambda) = \frac{x^k \lambda^{k-1}}{\Gamma(k)} \exp(-\lambda x),\qquad(3.12)$$

where $\Gamma(k)$ is the gamma function. Thus

$$M_2(\theta) = \int_{-\infty}^{\infty} \exp(\lambda\theta) P_2(\lambda) d\lambda = \int_0^{\infty} \exp(\lambda\theta) \frac{x^k \lambda^{k-1}}{\Gamma(k)} \exp(-\lambda x) d\lambda,$$

and, if $y = \lambda x$, then $dy = x d\lambda$, and

$$M_2(\theta) = \int_0^{\infty} \frac{\exp(\theta y/x)}{\Gamma(k)} y^{k-1} \exp(-y) dy$$

$$= \int_0^{\infty} \frac{y^{k-1}}{\Gamma(k)} \exp\left[-y\left(1 - \frac{\theta}{x}\right)\right] dy$$

$$= \left(1 - \frac{\theta}{x}\right)^{-k} \int_0^{\infty} \left[\left(1 - \frac{\theta}{x}\right)^k \bigg/ \Gamma(k)\right] y^{k-1} \exp\left[-y\left(1 - \frac{\theta}{x}\right)\right] dy$$

$$= \left(1 - \frac{\theta}{x}\right)^{-k} = \left(\frac{x}{x - \theta}\right)^k.\qquad(3.13)$$

Whence

$$G(s) = M_2[a(s-1)] = \left[\frac{x}{x - a(s-1)}\right]^k$$

$$= \left(\frac{x}{x + a - as}\right)^k = \left(1 + \frac{a}{x} - \frac{a}{x}s\right)^{-k}$$

$$= (1 + p - ps)^{-k} \qquad \left(p = \frac{a}{x}\right),\qquad(3.14)$$

which we recognize as the p.g.f. of the negative binomial distribution with density function

$$P(r) = \binom{k+r-1}{r} Q^r P^k,\qquad(3.15)$$

where

$$P = \frac{x}{x+a} = \frac{1}{1+p} \quad \text{and} \quad Q = 1 - P = \frac{a}{x+a} = \frac{p}{1+p}.$$

3.4 Generalized Poisson distributions

Let R_1 be a Poisson distributed random variable with mean v. Then the random variable $R \equiv R_1 \vee R_2$, with the probability density function in equation (3.3), is called a generalized Poisson random variable, and its distribution is referred to as a *generalized Poisson distribution*. Whence

$$G_1(s) = \sum_{r_1=0}^{\infty} \exp(-v) \frac{v^{r_1}}{r_1!} s^{r_1} = \exp[v(s-1)], \qquad (3.16)$$

and

$$G(s) = G_1[G_2(s)] = \exp\{v[G_2(s) - 1]\}. \qquad (3.17)$$

3.4.1 The Neyman Type A distribution

The Neyman Type A distribution can also arise as a generalized Poisson distribution in which the generalizer is another Poisson variable. Thus if

$$P_2(r_2) = \exp(-a) \frac{a^{r_2}}{r_2!},$$

with p.g.f.

$$G_2(s) = \exp[a(s-1)],$$

then

$$G(s) = \exp\langle v\{\exp[a(s-1)] - 1\}\rangle, \qquad (3.18)$$

and we have once again the Neyman Type A distribution of equation (3.11).

3.4.2 The negative binomial distribution

The negative binomial distribution can also be derived as a generalized Poisson distribution in which the generalizer is the logarithmic distribution. Thus if

$$P_2(r_2) = \frac{bQ^{r_2}}{r_2} \qquad (r_2 = 1, 2, \ldots), \ (0 < Q < 1), \qquad (3.19)$$

then, since

$$\sum_{r_2=1}^{\infty} P_2(r_2) = 1,$$

and

$$\sum_{r_2=1}^{\infty} \frac{Q^{r_2}}{r_2} = -\ln(1-Q),$$

we have that
$$b = -\frac{1}{\ln(1-Q)},$$
where $\ln(y)$ denotes the natural logarithm of y. Hence we find that
$$G_2(s) = \sum_{r_2=1}^{\infty} \frac{bQ^{r_2}s^{r_2}}{r_2} = \frac{\ln(1-Qs)}{\ln(1-Q)} - b\ln(1-Qs), \qquad (3.20)$$
and, by equation (3.17),
$$G(s) = \exp\{v[-b\ln(1-Qs) - 1]\}$$
$$= \exp[-vb\ln(1-Qs) - v]$$
$$= \exp[\ln(1-Qs)^{-vb} - v]$$
$$= (1-Qs)^{-vb}\exp(-v).$$
Let $k = vb$, then $v = k/b$ and
$$G(s) = (1-Qs)^{-k}\exp(-k/b)$$
$$= (1-Qs)^{-k}\exp[k\ln(1-Q)],$$
since $b = -[\ln(1-Q)]^{-1}$. Thus
$$G(s) = (1-Qs)^{-k}\exp[\ln(1-Q)^k] = (1-Qs)^{-k}(1-Q)^k$$
$$= \left(\frac{1-Q}{1-Qs}\right)^k = \left(\frac{P}{1-Qs}\right)^k, \qquad (3.21)$$
or, setting $Q = p/(1+p)$, we have that
$$G(s) = (1+p-ps)^{-k}. \qquad (3.22)$$
According to equation (2.18) this is the p.g.f. of the negative binomial distribution defined in equation (3.15).

3.5 Other compound and generalized distributions

The above procedures for generating the Neyman Type A and the negative binomial distributions as compound and generalized Poisson distributions can be carried out with many different combinations of distributions to create other new discrete probability distributions. We may, for example, assume other distributions for the compounding or generalizing variable, and, by substituting the m.g.f. or p.g.f. of this distribution into equation (3.8) or equation (3.17) respectively, generate other compound and generalized Poisson distributions. Or alternatively we may wish to deal with compound and generalized distributions that are not Poisson based. The Poisson-binomial and the Poisson-negative binomial distributions exemplify both sets of procedures.

3.5.1 The Poisson-binomial distribution
Let R_1 be a Poisson distributed random variable with mean v, and R_2 be a binomial distributed random variable with parameters n and p. Then, by equation (3.17), the generalized Poisson distribution of the random variable $R \equiv R_1 \vee R_2$ has the p.g.f.

$$G(s) = \exp\{v[G_2(s) - 1]\}$$
$$= \exp\{v[(1-p+ps)^n - 1]\}, \qquad (3.23)$$

which is the p.g.f. of the *Poisson-binomial distribution* (Neyman, 1939; Skellam, 1952)

$$P(r) = \exp(-v) \sum_{i=0}^{\infty} \frac{v^i}{i!} \binom{ni}{r} p^r (1-p)^{ni-r}, \qquad (3.24)$$

with mean and variance:

$$E(r) = m_1 = vnp,$$
$$\mathrm{var}(r) = m_2 = vnp[1 + (n-1)p].$$

Now let R_1 be a binomial distributed random variable with parameters nr_2 and p, and R_2 be a Poisson distributed random variable with mean v. Then, by equation (3.5), the *compound binomial distribution* of the random variable $R \equiv R_1 \wedge R_2$ has the p.g.f.

$$G(s) = \int_{-\infty}^{\infty} G_1(s|cr_2) P_2(r_2) dr_2$$
$$= \int_{-\infty}^{\infty} (1-p+ps)^{cr_2} P_2(r_2) dr_2$$
$$= \sum_{r_2=0}^{\infty} (1-p+ps)^{nr_2} \exp(-v) \frac{v^{r_2}}{r_2!}$$
$$= \exp\{v[(1-p+ps)^n - 1]\}, \qquad (3.25)$$

which is identical to the p.g.f. in equation (3.23). Consequently we conclude that a generalized Poisson distribution that is generalized by a binomial is *equivalent* to a compound binomial that is compounded by a Poisson. Or, in Gurland's (1957) notation:

Poisson $(v) \vee$ binomial $(n, p) \sim$ binomial $(nr_2, p) \wedge$ Poisson (v).

We shall see below that a similar equivalence exists when we replace the positive binomial with a negative binomial. Indeed Gurland (1957) demonstrates that this kind of equivalence holds for a large class of combinations.

3.5.2 The Poisson-negative binomial distribution
Let R_1 once again be a Poisson distributed random variable with mean v. However, now let us assume that R_2 is a negative binomial distributed

random variable with parameters k and p. Then the generalized Poisson distribution of $R \equiv R_1 \vee R_2$ has the p.g.f.

$$G(s) = \exp\{v[G_2(s) - 1]\}$$
$$= \exp\{v[(1 + p - ps)^k - 1]\}, \qquad (3.26)$$

which is the p.g.f. of the *Poisson–negative binomial distribution* (Skellam, 1952; Katti and Gurland, 1961)

$$P(r) = \exp(-v) \sum_{i=0}^{\infty} \frac{v^i}{i!} \binom{ki + r - 1}{r} p^r (1+p)^{-ki-r}, \qquad (3.27)$$

with mean and variance:

$E(r) = m_1 = vkp$,

$\text{var}(r) = m_2 = vkp[1 + (k+1)p]$.

Now let R_1 be a negative binomial distributed random variable with parameters kr_2 and p, and R_2 be a Poisson distributed random variable with mean v. Then the *compound negative binomial distribution* of the random variable $R \equiv R_1 \wedge R_2$ has the p.g.f.

$$G(s) = \int_{-\infty}^{\infty} G_1(s | cr_2) P_2(r_2) dr_2$$
$$= \int_{-\infty}^{\infty} (1 + p - ps)^{-cr_2} P_2(r_2) dr_2$$
$$= \sum_{r_2=0}^{\infty} (1 + p - ps)^{-kr_2} \exp(-v) \frac{v^{r_2}}{r_2!}$$
$$= \exp\{v[(1 + p - ps)^{-k} - 1]\}, \qquad (3.28)$$

which is identical to the p.g.f. in equation (3.26). We conclude, therefore, that

Poisson $(v) \vee$ negative binomial $(k, p) \sim$ negative binomial (kr_2, p) $\wedge$ Poisson (v),

to which we also may add our earlier conclusions that

Poisson $(v) \vee$ logarithmic $(Q) \sim$ Poisson $(\lambda a) \wedge$ gamma (v, k)

and

Poisson $(v) \vee$ Poisson $(a) \sim$ Poisson $(\lambda a) \wedge$ Poisson (v).

3.6 Some properties of the derived compound and generalized distributions
Table 3.1 summarizes our set of functions and measures for the component, compound, and generalized distributions that were derived in this and the preceding chapter. We shall now examine and compare some of their fundamental properties.

Compound and generalized distributions

Table 3.1. A summary of functions and measures for various component, compound, and generalized distributions.

Distribution	Probability generating function (p.g.f.) $G(s)$		Probability density function Probability $(R = r)$ $P(r)$	Mean $E(r)$	Variance $\mathrm{var}(r)$
Poisson	$\exp[v(s-1)]$,	$(v > 0)$	$\exp(-v)\dfrac{v^r}{r!}$	v	v
Binomial	$(1-p+ps)^n$,	$\begin{array}{l}(0 < p < 1)\\(n > 0)\end{array}$	$\dbinom{n}{r}p^r(1-p)^{n-r}$	np	$np(1-p)$
Negative binomial	$(1+p-ps)^{-k}$	$\begin{array}{l}(p > 0)\\(k > 0)\end{array}$	$\dbinom{k+r-1}{r}p^r(1+p)^{-k-r}$	$kp = w$	$kp(1+p) = w + \dfrac{w^2}{k}$
Neyman Type A	$\exp\{v\{\exp[a(s-1)]-1\}\}$	$(v, a > 0)$	$\exp(-v)\dfrac{a^r}{r!}\sum\limits_{i=0}^{\infty}\dfrac{t^i}{i!}[v\exp(-a)]^i (0^0 \equiv 1)$	va	$va(a+1)$
Poisson–binomial	$\exp\{v[(1-p+ps)^n - 1]\}$	$\begin{array}{l}(0 < p < 1)\\(n > 0)\end{array}$	$\exp(-v)\sum\limits_{i=0}^{\infty}\dfrac{v^i}{i!}\dbinom{ni}{r}p^r(1-p)^{ni-r}$	vnp	$vnp[1+(n-1)p]$
Poisson–negative binomial	$\exp\{v[(1+p-ps)^{-k} - 1]\}$	$\begin{array}{l}(p > 0)\\(k > 0)\end{array}$	$\exp(-v)\sum\limits_{i=0}^{\infty}\dfrac{v^i}{i!}\dbinom{ki+r-1}{r}p^r(1+p)^{-ki-r}$	vkp	$vkp[1+(k+1)p]$

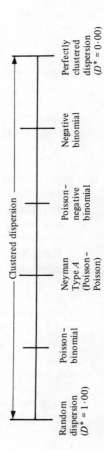

Figure 3.1. The dispersion line for clustered distributions.

In chapter 2, we demonstrated that both the binomial and the negative binomial distributions tended towards the Poisson, in the limit, under certain restrictive conditions. For the binomial distribution these conditions were that $n \to \infty$ and $p \to 0$, while the mean np remains constant and equal to λa, say. In the case of the negative binomial distribution, the corresponding restrictions were that $k \to \infty$ and $p \to 0$, while the mean $kp = \lambda a$.

The Poisson limiting form is also assumed by the Neyman Type A, the Poisson-binomial, and the Poisson-negative binomial distributions, when similar restrictions are placed on the limiting process. The Neyman Type A distribution tends to the Poisson as $v \to \infty$ and $a \to 0$, while the mean va remains constant and equal to λ, say. The Poisson-binomial distribution tends to the Poisson as $v \to \infty$, $p \to 0$, and the mean $vnp = \lambda$. Finally, the Poisson-negative binomial distribution assumes the Poisson form as $v \to \infty$, $p \to 0$, and the mean $vkp = \lambda$.

Some of these distributions also tend, in the limit, towards each other when their parameters take on extreme values. For example, the Poisson-binomial distribution very quickly assumes the form of the Neyman Type A distribution as n is increased. The Poisson-negative binomial tends to the Neyman Type A as $k \to \infty$ and $p \to 0$, while $pk = a$; it tends to the negative binomial, however, as $k \to 0$ and $v \to \infty$, while $vk = k_1$, say.

Of all the distributions in table 3.1, only the binomial has a variance-mean ratio that is less than unity, and therefore it is our only model of *regular* spatial dispersion. The Poisson distribution, we have seen, describes *random* spatial dispersion. The four remaining distributions all have a variance-mean ratio that is greater than unity and consequently describe *clustered* spatial dispersion. Which of these distributions are 'more clustered' than the others? Our answer depends, of course, on what we mean by 'more clustered'. In this book we shall take it to mean something quite specific. *A distribution will be said to be more clustered than another if, for a fixed mean and variance, it is more skewed than the other.*

Katti and Gurland (1961) evaluated the relative skewness of the negative binomial, the Neyman Type A, the Poisson-binomial, and the Poisson-negative binomial distributions for fixed mean $k_1 p_1$ and variance $k_1 p_1 (1+p_1)$, using an index of relative skewness. The numerical values of this index for these four distributions were found to be 2, 1, (0, 1), and (1, 2), respectively. Evidently the Poisson-negative binomial distribution spans the entire range of distributions from the Neyman Type A to the negative binomial, and the Poisson-binomial distribution covers the corresponding range between the Poisson and the Neyman Type A. Schematically, we may represent these ranges with a dispersion line on which we locate the distributions between the extremes of the perfectly regular and perfectly clustered dispersions that were defined in chapter 1. The right half of such a dispersion line appears in figure 3.1.

4
Parameter estimation

4.1 Introduction
Quadrat analysis has both a deductive and an inductive theory. The former strives to determine what frequency distribution tends to be generated by a particular theoretical stochastic process; the latter seeks to identify the stochastic process that is most likely to have generated an observed frequency distribution. The deductive theory has been considered in chapters 2 and 3. This and the following chapter focus on the inductive theory, and examine the related problems of parameter estimation and hypothesis testing.

4.2 Moment and maximum likelihood estimation
In order to carry out statistical tests on the quality of the fit provided by a theoretical model to an empirical distribution, it is first necessary to estimate the theoretical model's parameters. This estimation can be carried out in several ways. The most common procedures are to utilize the *method of moments* or the computationally more complex *method of maximum likelihood*. [Other methods such as the *method of sample zero frequency* and the *minimum chi-squared method* (Katti and Gurland, 1962b) have been used, and for certain regions in the parameter space these are frequently more efficient than moment estimators.]

To estimate the parameters of a distribution by the method of moments one simply equates the sample moments to their theoretical counterparts, or, more commonly, one uses the unbiased estimators of sample statistics such as the sample mean and sample variance.

When such moments exist in the theoretical distribution, this method provides *consistent* estimators of that distribution's parameters. However, several factors may be cited against the use of moment estimators. First, occasionally they do not exist. Second, moment estimators are seldom *efficient* estimators and may lead to erroneous conclusions regarding parametric interrelationships. Finally, the use of the chi-square goodness of fit test is justified only if efficient estimators are used (Chernoff and Lehmann, 1954).

Maximum likelihood estimators almost always exist, although they are sometimes difficult to obtain, because occasionally the numerical algorithms used to find them do not converge. Moreover, maximum likelihood estimators are efficient estimators if such estimators exist, and they are always asymptotically efficient.

4.2.1 Moment estimators
To derive the moments of a theoretical distribution, recall that, given the distribution's probability generating function (p.g.f.)

$$G(s) = \sum_{r=0}^{\infty} P(r)s^r ,$$

the following relationships hold:

$$G'(1) = \frac{d}{ds}G(s)\bigg|_{s=1} = \sum_{r=0}^{\infty} rP(r) = m_1 ,$$

$$G''(1) = \frac{d^2}{ds^2}G(s)\bigg|_{s=1} = \sum_{r=0}^{\infty} r(r-1)P(r) = m_2' - m_1 = m_2 + m_1^2 - m_1 ,$$

$$G'''(1) = \frac{d^3}{ds^3}G(s)\bigg|_{s=1} = \sum_{r=0}^{\infty} r(r-1)(r-2)P(r) = m_3' - 3m_2' + 2m_1 ,$$

where m_1 is the mean of the theoretical distribution, and m_i' and m_i are respectively the ith moment about zero and the ith moment about the mean. Note that for notational convenience we have defined m_1 to be m_1'.

Consequently,

$$m_1 = G'(1) ,$$
$$m_2 = G''(1) - [G'(1)]^2 + G'(1) , \qquad (4.1)$$
$$m_3 = G'''(1) - 3G'(1)G''(1) + 2[G'(1)]^3 + 3G''(1) - 3[G'(1)]^2 + G'(1) .$$

If, for example, k_1, k_2, and k_3 are the unknown parameters of the theoretical distribution, they are solutions of the following equations:

$$\hat{m}_1(k_1, k_2, k_3) = u_1 = \hat{m}_1 ,$$

$$\hat{m}_2(k_1, k_2, k_3) = u_2 \frac{N}{N-1} = \hat{m}_2 ,$$

$$\hat{m}_3(k_1, k_2, k_3) = u_3 \frac{N}{N-3+2/N} = \hat{m}_3 ,$$

where u_1 is the sample mean [8],

$$u_1 = \frac{1}{N}\sum_r rf_r , \qquad N = \sum_r f_r ,$$

(f_r is the number of observations in the rth frequency class), u_2 is the (biased) sample variance,

$$u_2 = u_2' - u_1^2 = \frac{1}{N}\sum_r (r^2 f_r) - u_1^2 ,$$

and u_3 is the third sample moment about the mean,

$$u_3 = u_3' - 3u_1 u_2' + 2u_1^3 = u_3' - 3u_1 u_2 - u_1^3 = \frac{1}{N}\sum_r (r^3 f_r) - 3u_1 u_2 - u_1^3 .$$

---

[8] For convenience, we shall henceforth adopt the notation $\sum_r$ to denote $\sum_{r=0}^{W}$, where W is the value of the largest observed frequency class.

4.2.2 Maximum likelihood estimators

Maximum likelihood estimators are those which maximize the likelihood function

$$L(k_1, k_2, ..., k_h) = \prod_{r=0}^{W} [P(r)]^{f_r},$$

where W is the largest frequency class observed (that is, $f_W \neq 0$ and $f_r = 0$ for all $r > W$).

To derive the maximum likelihood estimators, we solve the set of equations:

$$\frac{\partial}{\partial k_1} \ln L = 0,$$

$$\frac{\partial}{\partial k_2} \ln L = 0,$$

$$\vdots$$

$$\frac{\partial}{\partial k_h} \ln L = 0. \tag{4.2}$$

Equations (4.2) can be written in the more convenient form

$$\frac{\partial}{\partial k_i} \left\{ \sum_r \ln [P(r)]^{f_r} \right\} = 0 \qquad (i = 1, 2, ..., h),$$

or

$$\sum_r \frac{f_r}{P(r)} \frac{\partial}{\partial k_i} P(r) = 0 \qquad (i = 1, 2, ..., h). \tag{4.3}$$

In most cases, the roots of equation (4.3) are difficult to derive, and an iterative procedure is therefore employed. Moment estimators are frequently used as the initial approximations in such methods.

4.3 The Poisson distribution
4.3.1 The moment estimator

The p.g.f. of a Poisson distribution with parameter v is

$$G(s) = \exp v(s-1).$$

Hence

$$G'(1) = m_1 = v,$$

and the moment estimator of v is $\hat{v} = \hat{m}_1$.

Since m_2 is also equal to v, $\hat{m}_2$ is an alternative estimator of v.

4.3.2 The maximum likelihood estimator
For a Poisson distribution,

$$P(r) = \exp(-v)\frac{v^r}{r!},$$

and

$$\frac{d}{dv}P(r) = \frac{r}{v}P(r) - P(r).$$

Hence, recalling equation (4.3), we have that

$$\sum_r \frac{f_r}{P(r)} \frac{d}{dv}P(r) = 0,$$

or

$$\sum_r f_r \left(\frac{r}{v} - 1\right) = 0,$$

and, by definition,

$$\hat{v} = \frac{\sum_r rf_r}{\sum_r f_r} = \frac{1}{N}\sum_r rf_r = \hat{m}_1.$$

The maximum likelihood method yields the same estimator as the method of moments. It can be shown that this estimator is an unbiased and *fully* efficient estimator of v. This is not true of the sample variance.

4.4 The binomial distribution
4.4.1 The moment estimator
The p.g.f. of a binomial distribution is

$$G(s) = (1 - p + ps)^n,$$

with n a positive integer which is not usually considered as a parameter, but as a datum that takes the value of W, the largest frequency class observed. Hence

$$G'(1) = m_1 = np,$$

and the moment estimator of p is

$$\hat{p} = \frac{\hat{m}_1}{n} = \frac{\hat{m}_1}{W}.$$

4.4.2 The maximum likelihood estimator
For a binomial distribution,

$$P(r) = \binom{n}{r} p^r (1-p)^{n-r},$$

and
$$\frac{d}{dp}P(r) = \frac{r}{p}P(r) - \frac{(n-r)}{1-p}P(r) = \frac{r(1-p)-(n-r)p}{p(1-p)}P(r) = \frac{r-np}{p(1-p)}P(r).$$

Thus, by equation (4.3), we obtain

$$\sum_r \frac{f_r}{P(r)} \frac{d}{dp}P(r) = 0,$$

or

$$\sum_r f_r(r-np) = 0,$$

and

$$np = \frac{\sum_r r f_r}{\sum_r f_r} = \hat{m}_1, \quad \text{or} \quad \hat{p} = \frac{\hat{m}_1}{n} = \frac{\hat{m}_1}{w}.$$

As in the Poisson distribution, the maximum likelihood estimator and the moment estimator are identical.

4.5 The negative binomial distribution
4.5.1 The moment estimators
The p.g.f. of a negative binomial distribution is

$$G(s) = (1+p-ps)^{-k}$$

where p and k are positive parameters. Frequently, p is replaced by another parameter, w, such that $w = pk$ [as in equation (2.17)]. Then

$$G(s) = \left(1 + \frac{w}{k} - \frac{ws}{k}\right)^{-k},$$

$$\frac{d}{ds}G(s) = w\left(1 + \frac{w}{k} - \frac{ws}{k}\right)^{-(k+1)},$$

$$G'(1) = w,$$

$$G''(1) = \frac{w^2(k+1)}{k} = w^2 + \frac{w^2}{k}.$$

Thus

$$m_1 = w,$$

$$m_2 = w^2 + \frac{w^2}{k} - w^2 + w = w + \frac{w^2}{k},$$

and the moment estimators are

$$\hat{w} = \hat{m}_1,$$

$$\hat{k} = \frac{\hat{m}_1^2}{\hat{m}_2 - \hat{m}_1}.$$

4.5.2 The maximum likelihood estimators

For a negative binomial distribution with parameters w and k,

$$P(r) = \frac{k(k+1)\ldots(k+r-1)}{r!}\left(\frac{k}{w+k}\right)^k\left(\frac{w}{w+k}\right)^r,$$

and

$$\frac{\partial}{\partial w}P(r) = \frac{r}{w}P(r) - \frac{r+k}{w+k}P(r) = \frac{r(w+k)-w(r+k)}{w(w+k)}P(r) = \frac{k}{w+k}\left(\frac{r}{w}-1\right)P(r).$$

Equation (4.3) for w is therefore

$$\sum_r \frac{f_r}{P(r)}\frac{\partial}{\partial w}P(r) = 0,$$

or

$$\frac{k}{w+k}\sum_r \left(\frac{r}{w}-1\right)f_r = 0,$$

and

$$\hat{w} = \frac{\sum_r rf_r}{\sum_r f_r} = \hat{m}_1, \tag{4.4}$$

which is the same result as found by the method of moments.

We have, however, a second parameter, k, and therefore must express the second likelihood equation in the form defined by equation (4.3)

$$\sum_r \frac{f_r}{P(r)}\frac{\partial}{\partial k}P(r) = 0.$$

But,

$$\frac{\partial}{\partial k}P(r) = \sum_{i=0}^{r-1}\frac{P(r)}{k+i} + P(r)\ln\left(\frac{k}{k+w}\right) + \left(1 - \frac{k+r}{w+k}\right)P(r).$$

Therefore,

$$\sum_r f_r \sum_{i=0}^{r-1}\left(\frac{1}{k+i}\right) + \ln\left(\frac{k}{k+w}\right)\sum_r f_r + \frac{1}{w+k}\sum_r f_r(w-r) = 0, \tag{4.5}$$

and by equation (4.4) the last term of equation (4.5) is zero. Rewriting the first term of equation (4.5) as

$$\sum_r f_r \sum_{i=0}^{r-1}\frac{1}{k+i} = f_1\left(\frac{1}{k}\right) + f_2\left(\frac{1}{k}+\frac{1}{k+1}\right) + \ldots + f_W\left(\frac{1}{k}+\frac{1}{k+1}+\ldots+\frac{1}{k+W-1}\right)$$

$$= \frac{1}{k}(f_1+f_2+\ldots+f_W) + \frac{1}{k+1}(f_2+f_3+\ldots+f_W) + \ldots + \frac{1}{k+W-1}f_W,$$

Parameter estimation

and denoting the sum of observations whose frequency count is greater than r by S_r, that is

$$S_r = f_{r+1} + f_{r+2} + \ldots + f_W = \sum_{i=r+1}^{W} f_i \quad (r = 0, 1, \ldots, W),$$

we obtain from equation (4.5)

$$\sum_{r=0}^{W-1} \frac{S_r}{k+r} + N \ln\left(\frac{k}{k+w}\right) = 0 = Q(k), \qquad (4.6)$$

and we define $Q(k)$ by this equation.

Equation (4.6) can only be solved by iterative methods such as the Newton–Raphson method[9]. This begins with an initial approximation of k, say k_1, and obtains an improved approximation, say k_2, by means of the expression

$$k_2 = k_1 - \frac{Q(k_1)}{Q'(k_1)},$$

where $Q'(k)$ is the derivative of $Q(k)$. Thus we have

$$Q'(k) = N\left(\frac{1}{k} - \frac{1}{k+w}\right) - \sum_{r=0}^{W-1} \frac{S_r}{(k+r)^2},$$

where a convenient first approximation of k is provided by the moment estimator

$$\hat{k} = \frac{\hat{m}_1^2}{\hat{m}_2 - \hat{m}_1}.$$

To illustrate the differences between results obtained using the two methods of estimation, we present in table 4.1 both the moment and the maximum likelihood parameter estimates of w and k, and the associated expected frequency distributions, for the simulated negative binomial point pattern in figure 2.1b.

4.5.3 Existence and efficiency of estimators

Moment estimators for the negative binomial distribution exist provided that the variance of the sample is larger than the mean. Their efficiency depends on the values of w and k. If this efficiency is sufficiently large for a given w and k, then the tedious computations called for by the maximum likelihood method may not be worthwhile.

Sichel (1951), and Katti and Gurland (1962a) computed this efficiency for various values of w and k. Their results are summarized in figure 4.1. Although we only know the moment estimates of w and k when entering the curve in figure 4.1, in practice it has been found that they almost

---

[9] For a discussion of the Newton–Raphson method see any text on numerical analysis, such as Pennington (1965).

always lead to a reasonable decision. Note that the efficiency of the moments estimator in the case of the example in table 4.1 is relatively high (efficiency $\simeq 0 \cdot 8$). This accounts for the similarity of the moments and maximum likelihood estimates of k.

Table 4.1. Observed and expected quadrat distributions of the simulated clustered distribution; moment (mom.) and maximum likelihood estimators (m.l.e.) and the negative binomial model.

Number of points per cell	Number of cells observed	N.B. (mom.) $\hat{w} = 0 \cdot 5200$ $\hat{k} = 0 \cdot 6781$	N.B. (m.l.e.) $\hat{w} = 0 \cdot 5200$ $\hat{k} = 0 \cdot 7352$
0	67	67·98	67·48
1	23	20·01	20·55
2	5	7·29	7·39
3	2	2·82	2·79
4	2	1·13⎱	1·08⎱
5+	1	0·78⎰	0·70⎰

Total number of cells = 100
Total number of points = 52
$X^2 =$ [2·05]^a [2·12]^a
$\hat{m}_1 = 0 \cdot 5200$; $\hat{m}_2 = 0 \cdot 9188$
$P_{0 \cdot 05} =$ 5·99 5·99
$\hat{m}_2/\hat{m}_1 = 1 \cdot 7669$

^{a} $[X^2] = X^2$ statistic computed with grouping $\geqslant 1$ instead of $\geqslant 5$.

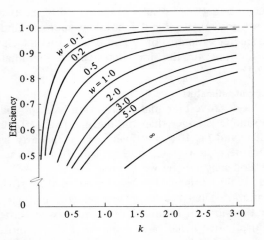

Figure 4.1. Efficiency of the method of moments estimator of k for a negative binomial distribution [adapted from Sichel (1951), and Katti and Gurland (1962a)].

4.6 The Neyman Type A distribution
4.6.1 The moment estimators
The p.g.f. of a Neyman Type A distribution with parameters v and a is

$$G(s) = \exp\langle v\{\exp[a(s-1)] - 1\}\rangle.$$

Hence

$$\frac{d}{ds}G(s) = va\exp[a(s-1)]G(s),$$

and

$$G'(1) = va, \qquad G''(1) = va(a+va).$$

Consequently

$$m_1 = va, \qquad m_2 = va(a+1),$$

and the moment estimators are

$$\hat{a} = \frac{\hat{m}_2 - \hat{m}_1}{\hat{m}_1}, \qquad \hat{v} = \frac{\hat{m}_1}{\hat{a}} = \frac{\hat{m}_1^2}{\hat{m}_2 - \hat{m}_1}.$$

4.6.2 The maximum likelihood estimators
For a Neyman Type A distribution,

$$P(r) = \exp(-v)\frac{a^r}{r!}\sum_{i=0}^{\infty}\frac{i^r}{i!}[v\exp(-a)]^i,$$

and

$$\begin{aligned}\frac{\partial}{\partial v}P(r) &= -P(r) + \exp(-v)\frac{a^r}{r!}\sum_{i=0}^{\infty}\frac{i^r}{i!}i\exp(-a)[v\exp(-a)]^{i-1} \\ &= -P(r) + \exp(-v)\frac{a^r}{r!}\sum_{i=0}^{\infty}\frac{i^{r+1}}{i!}\frac{[v\exp(-a)]^i}{v} \\ &= -P(r) + \frac{r+1}{va}\exp(-v)\frac{a^{r+1}}{(r+1)!}\sum_{i=0}^{\infty}\frac{i^{r+1}}{i!}[v\exp(-a)]^i \\ &= -P(r) + \frac{r+1}{va}P(r+1).\end{aligned}$$

Similarly,

$$\begin{aligned}\frac{\partial}{\partial a}P(r) &= \frac{r}{a}P(r) + \exp(-v)\frac{a^r}{r!}\sum_{i=0}^{\infty}\frac{i^r}{i!}i[-v\exp(-a)][v\exp(-a)]^{i-1} \\ &= \frac{r}{a}P(r) - \frac{r+1}{a}P(r+1).\end{aligned}$$

Therefore equation (4.3) may be expressed by the two likelihood equations:

$$\sum_r \frac{f_r}{P(r)} \frac{\partial}{\partial v} P(r) = \sum_r f_r \left[-1 + \frac{r+1}{va} \frac{P(r+1)}{P(r)} \right]$$

$$= -N + \frac{1}{va} \sum_r f_r(r+1) \frac{P(r+1)}{P(r)} = 0, \qquad (4.7)$$

$$\sum_r \frac{f_r}{P(r)} \frac{\partial}{\partial a} P(r) = \sum_r f_r \left[\frac{r}{a} - \frac{r+1}{a} \frac{P(r+1)}{P(r)} \right]$$

$$= \frac{1}{a} \sum_r r f_r - \frac{1}{a} \sum_r f_r(r+1) \frac{P(r+1)}{P(r)} = 0. \qquad (4.8)$$

If we let $H_r = (r+1)P(r+1)/P(r)$ and substitute in equation (4.8) the value of $\sum_r f_r H_r$ taken from equation (4.7), we obtain

$$\frac{1}{a} \sum_r r f_r - Nv = 0,$$

or

$$va = \frac{1}{N} \sum_r r f_r = \hat{m}_1.$$

The maximum likelihood estimators are therefore solutions of the following equations:

$$va = \hat{m}_1, \qquad (4.9)$$

$$\sum_r f_r H_r - N\hat{m}_1 = 0. \qquad (4.10)$$

Note that equation (4.9) was also the first equation obtained by the method of moments.

In equation (4.10), we can replace v by its value as found in equation (4.9) and write

$$Q(a) = \sum_r f_r H_r(a) - N\hat{m}_1 = 0, \qquad (4.11)$$

where $H_r(a)$ is the value of H_r when $v = \hat{m}_1/a$.

As with equation (4.6), the above equation can be solved by the Newton–Raphson method using the moment estimator of a as the first trial value. We obtain the derivative of $Q(a)$ as follows:

$$\frac{d}{da} P(r) = \frac{\partial}{\partial a} P(r) + \frac{\partial}{\partial v}[P(r)] \frac{dv}{da} = \frac{\partial}{\partial a} P(r) - \frac{\hat{m}_1}{a^2} \frac{\partial}{\partial v} P(r),$$

where v has been replaced by $\hat{m}_1/a$, and

$$\frac{\mathrm{d}}{\mathrm{d}a}P(r) = \frac{rP(r)-(r+1)P(r+1)}{a} + \frac{\hat{m}_1 P(r)-(r+1)P(r+1)}{a^2} . \qquad (4.12)$$

It can easily be shown, using equation (4.12), that

$$\frac{\mathrm{d}}{\mathrm{d}a}H_r = \frac{r+1}{a}\frac{P(r+1)}{P(r)} + \frac{a+1}{a^2}\left[\frac{(r+1)^2 P^2(r+1)}{P^2(r)} - \frac{(r+1)(r+2)P(r+2)}{P(r)}\right]$$

$$= H_r\left[\frac{1}{a} - \frac{a+1}{a^2}(H_{r+1}-H_r)\right] .$$

Therefore,

$$Q'(a) = \sum_r f_r \frac{\mathrm{d}}{\mathrm{d}a} H_r = \sum_r f_r H_r \left[\frac{1}{a} - \frac{a+1}{a^2}(H_{r+1}-H_r)\right] . \qquad (4.13)$$

Table 4.2 contrasts the moment and maximum likelihood estimates of the Neyman Type A distribution for the simulated clustered point pattern in figure 2.1b.

Table 4.2. Observed and expected quadrat distributions of the simulated clustered distribution; moment and maximum likelihood estimators and the Neyman Type A (N.T.A.) model.

Number of points per cell	Number of cells observed	N.T.A. (mom.) $\hat{v} = 0.6781$ $\hat{a} = 0.7669$	N.T.A. (m.l.e.) $\hat{v} = 0.9791$ $\hat{a} = 0.5311$
0	67	69.55	66.80
1	23	16.80	20.42
2	5	8.47	8.55
3	2	3.37 ⎫	2.94
4	2	1.21 ⎬	0.92 ⎫
5+	1	0.61 ⎭	0.37 ⎭
Total number of cells = 100			
Total number of points = 52			
$X^2 =$		3.81	[4.36][a]
$\hat{m}_1 = 0.5200$; $\hat{m}_2 = 0.9188$			
$P_{0.05} =$		3.84	5.99
$\hat{m}_2/\hat{m}_1 = 1.7669$			

[a] $[X^2] = X^2$ statistic computed with grouping ≥ 1 instead of ≥ 5.

4.6.3 Existence and efficiency of estimators

As for the negative binomial distribution, moment estimators exist when the variance of the sample is larger than its mean. Shenton (1949), and Katti and Gurland (1962a) calculated the efficiency of the moment estimators for different values of v and a. Their results are summarized in figure 4.2, which can be used to determine whether it is worthwhile to carry out the more laborious method of maximum likelihood once the moment estimators have been found. For our numerical example, the efficiency is somewhat lower (efficiency $\simeq 0 \cdot 6$) than in the case of the negative binomial model. This fact is reflected in the considerable differences between the moment and maximum likelihood estimates of v and a.

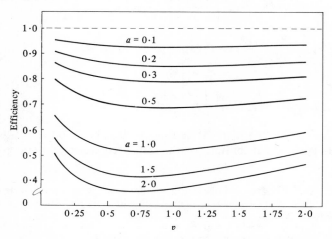

Figure 4.2. Efficiency of the method of moments for a Neyman Type A distribution [adapted from Shenton (1949), and Katti and Gurland (1962a)].

4.7 The Poisson-binomial distribution
4.7.1 The moment estimators
The p.g.f. of a Poisson-binomial distribution is

$$G(s) = \exp\{v[(1-p+ps)^n - 1]\},$$

where n is a positive integer. Because this distribution rapidly converges to a Neyman Type A distribution as n is increased, and since n is an integer, most applications of the Poisson-binomial have assumed n to be a datum and not an unknown parameter to be estimated[10]. Typically n is assumed to be equal to 2 or 3.

We therefore have that

$$\frac{d}{ds}G(s) = vnp(1-p+ps)^{n-1}G(s),$$

$$G'(1) = vnp,$$

$$G''(1) = vnp[(n-1)p + vnp].$$

---

[10] As will be seen later, the same assumption is not generally adopted for the Poisson-negative binomial distribution.

Parameter estimation

Hence
$$m_1 = vnp, \tag{4.14}$$
$$m_2 = vnp[1 + (n-1)p], \tag{4.15}$$

and the moment estimators are
$$\hat{p} = \frac{\hat{m}_2 - \hat{m}_1}{(n-1)\hat{m}_1}, \tag{4.16}$$
$$\hat{v} = \frac{\hat{m}_1}{n\hat{p}} = \frac{n-1}{n} \frac{\hat{m}_1^2}{\hat{m}_2 - \hat{m}_1}.$$

4.7.2 The maximum likelihood estimators
For a Poisson–binomial distribution,
$$P(r) = \exp(-v) \sum_{i=0}^{\infty} \frac{v^i}{i!} \binom{ni}{r} p^r (1-p)^{ni-r},$$
where
$$\binom{ni}{r} = \frac{(ni)!}{(ni-r)!r!}.$$

In the following we shall need to use the identity
$$(ni-r)\binom{ni}{r} = \frac{(ni)!(ni-r)}{(ni-r)!r!} = \frac{(ni)!(r+1)}{(ni-r-1)!(r+1)!} = (r+1)\binom{ni}{r+1}. \tag{4.17}$$

We begin by computing the partial derivatives of $P(r)$:
$$\frac{\partial}{\partial v} P(r) = -P(r) + \exp(-v) \sum_{i=0}^{\infty} \frac{iv^{i-1}}{i!} \binom{ni}{r} p^r (1-p)^{ni-r}$$
$$= -P(r) + \exp(-v) \sum_{i=0}^{\infty} \frac{v^i}{i!} \frac{(ni-r)+r}{n} \binom{ni}{r} p^r (1-p)^{ni-r}.$$

Use of the identity in equation (4.17) yields
$$\frac{\partial}{\partial v} P(r) = -P(r) + \frac{\exp(-v)}{vn} (r+1) \frac{1-p}{p} \sum_{i=0}^{\infty} \frac{v^i}{i!} \binom{ni}{r+1} p^{r+1} (1-p)^{ni-r-1}$$
$$+ \frac{r \exp(-v)}{vn} \sum_{i=0}^{\infty} \frac{v^i}{i!} \binom{ni}{r} p^r (1-p)^{ni-r}$$
$$= -P(r) + \frac{(r+1)(1-p)}{vnp} P(r+1) + \frac{r}{vn} P(r). \tag{4.18}$$

Similarly
$$\frac{\partial}{\partial p} P(r) = \exp(-v) \sum_{i=0}^{\infty} \frac{v^i}{i!} \binom{ni}{r} \left[\frac{r}{p} - \frac{ni-r}{(1-p)} \right] p^r (1-p)^{ni-r}$$
$$= \frac{r}{p} P(r) - \frac{r+1}{p} P(r+1). \tag{4.19}$$

The likelihood equations (4.3) can therefore be expressed as follows:

$$\sum_r \frac{f_r}{P(r)} \frac{\partial}{\partial v} P(r) = -\sum_r f_r + \frac{(1-p)}{vnp} \sum_r f_r(r+1) \frac{P(r+1)}{P(r)} + \frac{1}{vn} \sum_r r f_r = 0 \;.$$

Let

$$H_r = (r+1) \frac{P(r+1)}{P(r)} \quad \text{and} \quad \sum_r r f_r = \hat{m}_1 \sum_r f_r = N\hat{m}_1 \;.$$

Then

$$\sum_r \frac{f_r}{P(r)} \frac{\partial}{\partial v} P(r) = -N + \frac{N\hat{m}_1}{vn} + \frac{(1-p)}{vnp} \sum_r f_r H_r = 0 \;. \tag{4.20}$$

Similarly

$$\sum_r \frac{f_r}{P(r)} \frac{\partial}{\partial p} P(r) = \frac{1}{p} \sum_r r f_r - \frac{1}{p} \sum_r f_r H_r = 0 \;,$$

or

$$\frac{N\hat{m}_1}{p} - \frac{1}{p} \sum_r f_r H_r = 0 \;,$$

which yields

$$\sum_r f_r H_r - N\hat{m}_1 = 0 \;.$$

Replacing, in equation (4.20), $\sum_r f_r H_r$ by $N\hat{m}_1$, we obtain

$$\sum_r \frac{f_r}{P(r)} \frac{\partial}{\partial v} P(r) = -N + \frac{N\hat{m}_1}{vn} + \frac{(1-p)}{vnp} N\hat{m}_1 = N\left(-1 + \frac{\hat{m}_1}{vnp}\right) = 0 \;.$$

Therefore the maximum likelihood estimators are solutions of the following system of equations:

$$vnp = \hat{m}_1 \;, \tag{4.21}$$

$$\sum_r f_r H_r - N\hat{m}_1 = 0 \;. \tag{4.22}$$

Note that equation (4.21) is the same as the first moment equation, and that equation (4.22) bears a similarity to equation (4.10). Writing $H_r(p)$ for the value of H_r when $v = \hat{m}_1/np$, we have

$$Q(p) = \sum_r f_r H_r(p) - N\hat{m}_1 = 0 \;, \tag{4.23}$$

and, as before, we may solve this equation by the Newton–Raphson iterative procedure. Thus

$$\frac{d}{dp} P(r) = \frac{\partial}{\partial p} P(r) + \frac{\partial}{\partial v} P(r) \frac{dv}{dp} = \frac{\partial}{\partial p} P(r) - \frac{\hat{m}_1}{np^2} \frac{\partial}{\partial v} P(r) \;,$$

which, by equations (4.18) and (4.19), becomes
$$\frac{d}{dp}P(r) = \frac{rP(r)-(r+1)P(r+1)}{p}$$
$$+\frac{\hat{m}_1}{np^2}\left[P(r)-\frac{r}{vn}P(r)-\frac{(r+1)(1-p)}{vnp}P(r+1)\right].$$

Replacing v by $\hat{m}_1/np$, and collecting terms in $P(r)$ and $P(r+1)$, we have
$$\frac{d}{dp}P(r) = P(r)\left[\frac{(n-1)}{n}\frac{r}{p}+\frac{\hat{m}_1}{np^2}\right] - (r+1)P(r+1)\left[\frac{(n-1)}{n}\frac{1}{p}+\frac{1}{np^2}\right]. \quad (4.24)$$

One can easily verify, using equation (4.24), that
$$\frac{d}{dp}H_r = (r+1)\frac{P(r+1)}{P(r)}\frac{n-1}{np} - \left(\frac{n-1}{np}+\frac{1}{np^2}\right)\left[\frac{(r+1)(r+2)P(r+2)}{P(r)}\right.$$
$$\left.-\frac{(r+1)^2P^2(r+1)}{P^2(r)}\right]$$
$$= H_r\left[\frac{n-1}{np} - \left(\frac{n-1}{np}+\frac{1}{np^2}\right)(H_{r+1}-H_r)\right].$$

Therefore
$$Q'(p) = \sum_r f_r \frac{d}{dp} H_r = \sum_r f_r H_r\left[\frac{n-1}{np} - \left(\frac{n-1}{np}+\frac{1}{np^2}\right)(H_{r+1}-H_r)\right]. \quad (4.25)$$

It can be shown that as $n \to \infty$ in such a way that $np \to a$, then equation (4.25) tends to equation (4.13).

4.7.3 Existence and efficiency of estimators

As previously mentioned, the moment estimators exist if and only if the variance u_2 lies between u_1 and $n\hat{m}_1$. Sprott (1958), and Katti and Gurland (1962b) have tabulated their efficiency for $n = 2, 3,$ and 5, for various values of p and v. Their results for $n = 2$ are summarized in figure 4.3, which shows that the efficiency of the method of moments is very low as soon as p is larger than 0·2. In fact for any value of v the efficiency tends to zero as $p \to 1$.

These authors observe that improved efficiency could be obtained by using the method of sample zero frequency—a method in which the observed proportion of zero counts is used to form one estimating equation, the other being fitted by the maximum likelihood equation, equation (4.21). We have then
$$P(0) = \frac{f_0}{N} = \exp\{-v[1-(1-p)^n]\}, \quad (4.26)$$

which yields when $n = 2$
$$\hat{p} = 2 + \frac{2}{\hat{m}_1}\ln\frac{f_0}{N} \quad \text{and} \quad \hat{v} = \frac{\hat{m}_1}{n\hat{p}}.$$

When n is larger than 2, equation (4.26) is not solved easily, but the Newton-Raphson method can always be used.

As the estimator $\hat{p}$ must lie between 0 and 1, the sample zero frequency estimators exist, provided that the following inequalities hold

$$N\exp(-\hat{m}_1) \leqslant f_0 \leqslant N\exp(-\hat{m}_1/n) \ .$$

Therefore it may happen that the moment estimators do not exist, whereas the sample zero frequency estimators do exist—and *vice versa*.

The efficiency of this method is much larger than that of the method of moments. This is shown for $n = 2$ in figure 4.4, which was derived using the results of Sprott (1958), and Katti and Gurland (1962b).

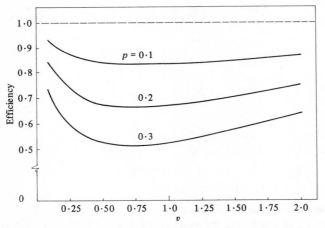

Figure 4.3. Efficiency of the method of moments for a Poisson-binomial distribution with $n = 2$ [adapted from Sprott (1958), and Katti and Gurland (1962b)].

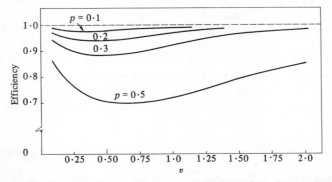

Figure 4.4. Efficiency of the method of sample zero frequency estimator of p for a Poisson-binomial distribution with $n = 2$ [adapted from Sprott (1958), and Katti and Gurland (1962b)].

Table 4.3 illustrates the different estimates of the Poisson–binomial's parameters that may be obtained with our numerical example. In addition to the model with $n = 2$, we also include the fit of the model with $n = 4$ in order to show the rapid convergence of the expected frequency distribution toward that of the Neyman Type A. Note that the sample zero frequency estimates lie between the moment and maximum likelihood estimates.

Table 4.3. Observed and expected quadrat distributions of the simulated clustered distribution; moment and maximum likelihood estimators and the Poisson–binomial (P.B.) model.

Number of points per cell	Number of cells observed	P.B. ($n=2$) (mom.) $\hat{v} = 0{\cdot}3390$ $\hat{p} = 0{\cdot}7669^{b}$	P.B. ($n=2$) (m.l.e.) $\hat{v} = 0{\cdot}7379$ $\hat{p} = 0{\cdot}3524$	P.B. ($n=4$) (m.l.e.) $\hat{v} = 0{\cdot}8438$ $\hat{p} = 0{\cdot}1541$
0	67	72·57	65·16	66·25
1	23	8·80	21·94	20·86
2	5	15·00	9·66	8·98
3	2	1·78	2·43 ⎫	2·83 ⎫
4	2	1·55 ⎫	0·65 ⎬	0·80 ⎬
5+	1	0·30 ⎭	0·17 ⎭	0·28 ⎭
Total number of cells = 100				
Total number of points = 52				
$X^2 =$		[30·77]^{a}	[3·31]^{a}	[5·63]^{a}
$\hat{m}_1 = 0{\cdot}5200; \hat{m}_2 = 0{\cdot}9188$				
$P_{0 \cdot 05} =$		5·99	3·84	5·99
$\hat{m}_2/\hat{m}_1 = 1{\cdot}7669$				

[a] $[X^2] = X^2$ statistic computed with grouping ≥ 1 instead of ≥ 5.
[b] Sample zero frequency estimates of $\hat{v}$ and $\hat{p}$ are 0·5656 and 0·4597, respectively.

4.8 The Poisson–negative binomial distribution
4.8.1 The moment estimators

The p.g.f. of a Poisson–negative binomial distribution is

$$G(s) = \exp\{v[(1+p-ps)^{-k} - 1]\},$$

where both p and k are positive. Formally it is the same as a Poisson–binomial distribution, except that p has been replaced by $-p$, and n by $-k$. Here, however, k is no longer an integer but is a third unknown parameter which is to be estimated from the data.

Proceeding as before, we compute

$$\frac{d}{ds}G(s) = vkp(1+p-ps)^{-k-1}G(s),$$

$$\frac{d^2}{ds^2}G(s) = vkp[(k+1)p(1+p-ps)^{-k-2} + vkp(1+p-ps)^{-2(k+1)}]G(s),$$

and obtain

$$G'(1) = vkp,$$
$$G''(1) = vkp[(k+1)p + vkp],$$
$$G'''(1) = vkp[(k+1)(k+2)p^2 + 2vkp(k+1)p + vkp(k+1)p + v^2k^2p^2].$$

Consequently, by equation (4.1),

$$m_1 = vkp, \tag{4.27}$$
$$m_2 = vkp[1 + (k+1)p], \tag{4.28}$$

or

$$p = \frac{m_2 - m_1}{(k+1)m_1}. \tag{4.29}$$

Note the correspondence between equations (4.27) and (4.28), and equations (4.14) and (4.15), respectively.

Finally, again by equation (4.1), we have that

$$m_3 = vkp[(k+1)(k+2)p^2 + 3vkp(k+1)p + v^2k^2p^2 - 3vkp(k+1)p$$
$$- 3v^2k^2p^2 + 2v^2k^2p^2 + 3(k+1)p + 3vkp - 3vkp + 1]$$
$$= vkp[(k+1)(k+2)p^2 + 3(k+1)p + 1].$$

Using equations (4.27) and (4.29), we obtain

$$m_3 = m_1 \left[\frac{(k+1)(k+2)(m_2 - m_1)^2}{(k+1)^2 m_1^2} + \frac{3(k+1)(m_2 - m_1)}{(k+1)m_1} + 1 \right]$$
$$= \frac{(m_2 - m_1)^2}{m_1} \frac{k+2}{k+1} + 3m_2 - 2m_1,$$

which can be written as

$$(k+1)m_1(m_3 - 3m_2 + 2m_1) = (m_2 - m_1)^2(k+2),$$

or

$$k = \frac{2(m_2 - m_1)^2 - (m_3 - 3m_2 + 2m_1)m_1}{m_1(m_3 - 3m_2 + 2m_1) - (m_2 - m_1)^2} = \frac{2m_2^2 - m_1(m_2 + m_3)}{m_1(m_1 - m_2 + m_3) - m_2^2}. \tag{4.30}$$

From equations (4.27), (4.29), and (4.30), we obtain the moment estimators:

$$\hat{k} = \frac{2\hat{m}_2^2 - \hat{m}_1(\hat{m}_2 + \hat{m}_3)}{\hat{m}_1(\hat{m}_1 - \hat{m}_2 + \hat{m}_3) - \hat{m}_2^2}, \tag{4.31}$$

$$\hat{p} = \frac{\hat{m}_2 - \hat{m}_1}{(\hat{k}+1)\hat{m}_1}, \tag{4.32}$$

$$\hat{v} = \frac{\hat{m}_1}{\hat{k}\hat{p}}. \tag{4.33}$$

4.8.2 The maximum likelihood estimators

For a Poisson–negative binomial distribution,

$$P(r) = \exp(-v) \sum_{i=0}^{\infty} \frac{v^i}{i!} \frac{ki(ki+1)\ldots(ki+r-1)}{r!} p^r (1+p)^{-ki-r}.$$

Thus, using the same process as for the Poisson–binomial distribution, we obtain

$$\frac{\partial}{\partial v}P(r) = -P(r) + \frac{(r+1)(1+p)}{vkp}P(r+1) - \frac{r}{vk}P(r),$$

$$\frac{\partial}{\partial p}P(r) = \frac{rP(r)-(r+1)P(r+1)}{p},$$

and the first two likelihood equations are

$$\sum_r \frac{f_r}{P(r)} \frac{\partial}{\partial v} P(r) = 0,$$

and

$$\sum_r \frac{f_r}{P(r)} \frac{\partial}{\partial p} P(r) = 0,$$

which reduce to

$$vkp = \hat{m}_1, \tag{4.34}$$

and

$$Q(p,k) = \sum_r f_r H_r(p,k) - N\hat{m}_1 = 0, \tag{4.35}$$

where

$$H_r = (r+1) \frac{P(r+1)}{P(r)}.$$

Holding k constant, we obtain

$$\frac{d}{dp} Q(p) = Q'(p) = \sum_r f_r H_r \left[\frac{k+1}{kp} - \left(\frac{k+1}{kp} + \frac{1}{kp^2} \right)(H_{r+1} - H_r) \right]. \tag{4.36}$$

To solve the third likelihood equation

$$\sum_r \frac{f_r}{P(r)} \frac{\partial}{\partial k} P(r) = 0, \tag{4.37}$$

we must first obtain $\partial P(r)/\partial k$. Shumway and Gurland (1960b) show, after laborious computation, that

$$\frac{\partial}{\partial k} P(r) = \frac{rP(r)}{k} - \frac{1}{k} \sum_{i=1}^{r-1} B_{ri} P(i) - \left[\frac{(1+p)(r+1)}{pk} P(r+1) - \frac{r}{k} P(r) \right] \ln(1+p),$$

where

$$B_{ri} = \left(\frac{p}{1+p}\right)^{r-i} \frac{i}{(r-i)(r-i+1)} .$$

Thus equation (4.37) becomes

$$\frac{1}{k}\sum_r rf_r - \frac{\ln(1+p)}{k}\left[\frac{1+p}{p}\sum_r f_r H_r - \sum_r rf_r\right] - \frac{1}{k}\sum_r \frac{f_r}{P(r)} \sum_{i=1}^{r-1} B_{ri} P(i) = 0 .$$

Recalling equation (4.35), and the fact that $\sum_r rf_r = N\hat{m}_1$, and multiplying by k, we obtain

$$N\hat{m}_1 - \ln(1+p)\left(\frac{1+p}{p} - 1\right) N\hat{m}_1 - \sum_r \frac{f_r}{P(r)} \sum_{i=1}^{r-1} B_{ri} P(i) = 0 ,$$

or

$$N\hat{m}_1\left[1 - \frac{\ln(1+p)}{p}\right] - \sum_r \frac{f_r}{P(r)} \sum_{i=1}^{r-1} B_{ri} P(i) = 0 .$$

Since $\hat{m}_1 = vkp$, the above equation reduces to

$$k = \frac{1}{Nv[p - \ln(1+p)]} \sum_r \frac{f_r}{P(r)} \sum_{i=1}^{r-1} B_{ri} P(i) . \tag{4.38}$$

We can now find the maximum likelihood estimators by the following iterative procedure:

1 Let v', k', p' be initial estimates, for example, the moment estimates obtained from equations (4.31), (4.32), and (4.33).
2 Using equations (4.35) and (4.36), compute a new estimate of p,

$$p'' = p' - \frac{Q(v', k', p')}{Q'(v', k', p')} .$$

3 Using equation (4.38), compute a new estimate of k,

$$k'' = \text{function of } (v', k', p'') .$$

4 Using equation (4.34), compute a new value of v,

$$v'' = \frac{\hat{m}_1}{k''p''} .$$

5 If v'', k'', p'' are significantly different from v', k', p', repeat steps 2, 3, and 4 with v'', k'', p'' as the new approximations.

4.8.3 Existence and efficiency of estimators

The moment estimators exist if equations (4.31), (4.32), and (4.33) yield positive values. This is the case if and only if the following inequalities hold:

$$\hat{m}_2 \geqslant \hat{m}_1 , \qquad \frac{\hat{m}_2^2 + \hat{m}_1 \hat{m}_2 - \hat{m}_1^2}{\hat{m}_1} \leqslant \hat{m}_3 \leqslant \frac{\hat{m}_2(2\hat{m}_2 - \hat{m}_1)}{\hat{m}_1} .$$

Katti and Gurland (1961) calculated the efficiency of the moment estimators for various values of v, k, and p; they are very low when p is larger than $0 \cdot 1$, or k is larger than 1. Apparently, much better results can be obtained by using the ratio of the first two observed frequencies, instead of the third moment, to obtain the third estimating equation. In that method the value of p is obtained first, as the solution of

$$\hat{p} \ln\left(\hat{m}_1 \frac{f_0}{f_1}\right) - \frac{\hat{m}_2 - \hat{m}_1}{\hat{m}_1} \ln(1+\hat{p}) = 0 ,$$

which can easily be solved by the Newton–Raphson method. The estimates $\hat{k}$ and $\hat{v}$ are obtained from equations (4.32) and (4.33). These estimators exist if and only if

$$1 + \ln\left(\hat{m}_1 \frac{f_0}{f_1}\right) \leqslant \frac{\hat{m}_2}{\hat{m}_1} \leqslant \hat{m}_1 \frac{f_0}{f_1} ,$$

and they provide better starting values than the moment estimators for the iterative process needed to obtain the maximum likelihood estimators.

However, in many instances, this iterative process does not converge. The reason is probably the iterative use of equation (4.38) which has the form

$$k = f(k) .$$

If we have an approximate value k_1 of the solution of the equation, then

$$k_2 = f(k_1)$$

is a better approximation if $df(k)/dk < 1$, in the interval $(k_1, \hat{k})$, as is shown in figure 4.5.

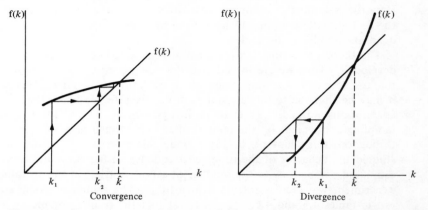

Figure 4.5. Convergence and divergence of the iterative process used to solve $k = f(k)$.

Table 4.4 presents the moment estimates of the Poisson-negative binomial model's parameters when they are fitted to our numerical example. The algorithms for the maximum likelihood method and the ratio of the first two observed frequencies method both diverged.

Table 4.4. Observed and expected quadrat distributions of the simulated clustered distribution; moment estimators and the Poisson-negative binomial (P.N.B.) model.

Number of points per cell	Number of cells observed	P.N.B. (mom.) $\hat{v} = 1\cdot7016$ $\hat{k} = 0\cdot6624$ $\hat{p} = 0\cdot4613$
0	67	68·52
1	23	18·96
2	5	7·60
3	2	3·01
4	2	1·18 ⎞
5+	1	0·73 ⎠
Total number of cells = 100		
Total number of points = 52		
$X^2 =$		[2·75][a]
$\hat{m}_1 = 0\cdot5200;\ \hat{m}_2 = 0\cdot9188$		
$P_{0\cdot05} =$		3·84
$\hat{m}_2/\hat{m}_1 = 1\cdot7669$		

[a] $[X^2] = X^2$ statistic computed with grouping $\geqslant 1$ instead of $\geqslant 5$.

4.9 Quadrat sampling and quadrat censusing

The theory of estimation that has been developed in this chapter is based on *quadrat sampling*—a procedure that selects quadrats at random in the study area. It is obvious that this theory is not exactly applicable to the case of *quadrat censusing*, which has been used in our numerical example. Here the 'sampling' was carried out with contiguous quadrats that covered the entire study area, in which case the numbers of points in adjacent quadrats are clearly dependent.

The theory of estimation for quadrat censusing has not yet been developed. Thus we are forced into the position of using procedures that are founded on assumptions which we know to be false. However, it may be that the results are not significantly affected when the assumptions are false. To examine this possibility, we conducted a small sampling experiment on the numerical example that has been used throughout this chapter. Taking a random sample of 25% each time, we studied the behavior of the observed frequency distribution and several estimated parameters, as the number of sampling trials in the sampling experiment was increased from five to fifty. The principal results are set out in tables 4.5 and 4.6. They suggest that the results obtained with quadrat sampling tend toward the results obtained by quadrat censusing as the number of sampling trials is increased.

Table 4.5. A comparison of observed frequency distributions obtained by quadrat sampling and by quadrat censusing.[a]

Number of points per cell	Number of cells observed with quadrat sampling				Number of cells observed with quadrat censusing
	Number of sampling trials				
	5	10	20	50	
0	16·0	16·5	16·35	16·70	16·75 (67)
1	4·8	5·3	5·70	5·72	5·75 (23)
2	2·6	1·9	1·70	1·42	1·25 (5)
3	0·8	0·6	0·50	0·40	0·50 (2)
4	0·6	0·5	0·45	0·42	0·50 (2)
5+	0·2	0·2	0·30	0·34	0·25 (1)

[a] The numbers in brackets are the corresponding figures for a 100% sample.

Table 4.6. Comparison of parameter estimates obtained by quadrat sampling and by quadrat censusing.

Parameters	Parameter estimates with quadrat sampling				Parameter estimates with quadrat censusing
	Number of sampling trials				
	5	10	20	50	
Mean	0·6320	0·5560	0·5560	0·5256	0·5200
Variance	1·0807	0·9430	0·9590	0·9326	0·9097
Negative binomial					
Moment estimates					
$\hat{w}$	0·6320	0·5560	0·5560	0·5256	0·5200
$\hat{k}$	0·8902	0·7989	0·7672	0·6787	0·6781
Maximum likelihood estimates					
$\hat{w}$	0·6320	0·5560	0·5560	0·5256	0·5200
$\hat{k}$	0·7329	0·7487	0·7821	0·7316	0·7352
Neyman Type A					
Moment estimates					
$\hat{v}$	0·8902	0·7989	0·7672	0·6787	0·6781
$\hat{a}$	0·7099	0·6960	0·7247	0·7744	0·7669
Maximum likelihood estimates					
$\hat{v}$	0·8551	0·9223	1·0026	0·9733	0·9791
$\hat{a}$	0·7391	0·6028	0·5545	0·5400	0·5311

5

Hypothesis testing: the chi-square goodness of fit test

5.1 Introduction
Probability theory strives to answer questions of the following kind: given that one tosses an unbiased coin, what is the probability of obtaining exactly 20 heads in 50 tosses? Statistical inference on the other hand addresses questions such as: given that exactly 20 heads have occurred in 50 tosses of a coin, what can one infer about the 'unbiasedness' of the coin? Thus, in quadrat analysis, we may distinguish between efforts which focus on the development of stochastic models of point dispersions, and efforts which deal with the problem of inference regarding the model that may be said to 'account' for an observed point dispersion. Efforts of the first kind address questions such as: given that 150 points are randomly placed inside a grid of 100 squares, what is the expected number of unoccupied squares, squares containing one point, and so on? Efforts of the second kind consider questions such as: given that in a grid of 100 squares, superimposed over a point pattern, 35 squares are empty, 30 squares contain a single point, and so on, what can one infer about the stochastic model that may have generated this point dispersion? This chapter deals with the latter class of problems and, in particular, focuses on the use of the chi-square goodness of fit test to measure the adequacy of fit, to an observed frequency distribution, that is provided by alternative quadrat models of point dispersion.

Let $x_1, x_2, ..., x_N$ denote independent observations on a random variable which has an unknown distribution function $F(x)$. The problem of testing whether $F(x) = F_0(x)$, where $F_0(x)$ is some particular distribution function, is called a goodness of fit problem, and any test of the null hypothesis

$$H_0: F(x) = F_0(x) , \tag{5.1}$$

is called a goodness of fit test.

Hypotheses of fit, like parametric hypotheses, may be divided into two classes: simple and composite. In equation (5.1) H_0 is a simple hypothesis if $F_0(x)$ is completely specified. Otherwise H_0 is a composite hypothesis. For example, the hypothesis that the N observations of a random sample have been generated by a Poisson spatial process with mean equal to λa is a simple hypothesis. The hypothesis that the N observations were produced by a Poisson spatial process whose parameter is unknown, however, is a composite hypothesis.

5.2 The chi-square goodness of fit test
The oldest and most commonly used method for evaluating the fit of a theoretical distribution to an observed one is the *chi-square goodness of fit test* (hereafter referred to simply as the chi-square test). For this test

the empirical data are first grouped into mutually exclusive discrete frequency classes and then compared with the expected frequency distribution that would arise from a particular theoretical model. This comparison leads to the calculation of a test statistic which approximately follows a chi-square distribution, *only if the hypothesized model is correct.* If the model is not the correct one, the test statistic will tend to exceed the chi-square variate. Thus an ordinary table of chi-square percentiles may be used to determine whether the model provides a satisfactory accounting for the data.

The computational procedure for carrying out the chi-square test begins with the insertion of the parameter values into the probability density function of the theoretical model, and the sequential computation of the probabilities that a random variable from the assumed distribution falls within each frequency class. Each of these probabilities when multiplied by the number of observations in the sample yields the expected number of observations which would fall into each frequency class under the assumed model. Generally, we adopt the rule of thumb that the expected number of observations in each class should always be at least five, and classes with lower expected observations are therefore grouped together. At this point we compute the following test statistic[11]:

$$X^2 = \sum_{r=0}^{W} \frac{[f_r - NP_0(r)]^2}{NP_0(r)} , \qquad (5.2)$$

where $W+1$ denotes the number of frequency classes, f_r denotes the number of empirical observations in each frequency class, N is the sample size $\left(\sum_{r=0}^{W} f_r = N\right)$, and $P_0(r)$ denotes the probability that an observation falls into the rth frequency class, under the null hypothesis. Finally, we compare the computed test statistic with tabulated percentiles for a chi-square distribution with the appropriate degrees of freedom. High values of the test statistic lead to a rejection of the null hypothesis.

The essence of the chi-square test of fit is the reduction of the problem to one involving the multinomial distribution. If the null hypothesis is correct, the observed expected values for each frequency class follow a multinomial distribution with the $P(r)$ as probabilities. The joint distribution of the f_r is therefore given by

$$\frac{N!}{f_0! f_1! f_2! \ldots f_W!} P_0(0)^{f_0} P_0(1)^{f_1} P_0(2)^{f_2} \ldots P_0(W)^{f_W} .$$

If the null hypothesis is simple, and the $P_0(r)$ are assumed to be known, we may draw on the well-known result that the multinomial distribution

---

[11] To avoid confusion we shall use the symbol X^2 to denote the chi-square statistic in equation (5.2), and the symbol χ^2 to denote the random variable which follows the chi-square distribution.

approaches the normal distribution for large N, and conclude that each

$$X_r = \frac{f_r - NP_0(r)}{[NP_0(r)]^{1/2}} \tag{5.3}$$

is asymptotically normal with zero mean and unit variance. Hence, as $N \to \infty$,

$$X^2 = \sum_{r=0}^{W} X_r^2$$

tends to the sum of the squares of $W+1$ independent normal variates, subject to the single homogeneous linear constraint [12]

$$\sum_{r=0}^{W} f_r = N,$$

or

$$\sum_{r=0}^{W} X_r [NP_0(r)]^{1/2} = 0.$$

It follows therefore that X^2, in the limit, is distributed as a χ^2 distribution with W degrees of freedom.

If the null hypothesis is a composite hypothesis (that is, if some or all of the parameters are left unspecified), a new dimension is interjected into the problem. The sampling distribution is altered because now the theoretical probabilities of the multinomial distribution are no longer assumed to be known, but are themselves random variables which are functions of the h unspecified parameters. It is not obvious that the asymptotic distribution of X^2 in this situation will be of the same form as in the case of the simple hypothesis. However, this is the case (Kendall and Stuart, 1961), and the only effect is that h additional homogeneous linear constraints are imposed on the f_r, thereby reducing the number of degrees of freedom to $W - h$.

The principal advantages of the chi-square test are its simplicity and versatility. It can easily be applied to test the fit of any probability distribution, and it does not require prior knowledge of the distribution's parameters. Its major drawback is its lack of sensitivity. And, as we shall see below, grouping the data can influence the outcome of the test.

Alternatives to the chi-square test are of two kinds. First we can identify those tests which, like the chi-square test, are 'general' tests. Second, we have available a class of 'specific' tests that can be used in situations where the alternative hypothesis is more definitely specified.

---

[12] By equation (5.3), $f_r = X_r[NP_0(r)]^{1/2} + NP_0(r)$,

therefore $\sum_{r=0}^{W} f_r = \sum_{r=0}^{W} X_r[NP_0(r)]^{1/2} + N\sum_{r=0}^{W} P_0(r) = 0 + N = N.$

Because the application of quadrat methods in urban analysis is still in its infancy, 'specific' tests have not yet received much attention. [A notable exception is the recent paper by Hinz and Gurland (1970).]

The 'general' test alternatives to the chi-square test, such as the likelihood ratio test, have not provided it with serious competition for a number of reasons. The most significant reason is that they do not offer advantages not already possessed by the chi-square test. Moreover a few of them, such as the Kolmogorov–Smirnov test, require a complete specification of the frequency distribution that is followed by the observations. Thus, despite the many problems associated with its use, the chi-square test is still the most useful test for goodness of fit that is currently available for quadrat analysis.

5.3 The power of the chi-square test

In the Neyman–Pearson theory of hypothesis testing, the efficacy of a statistical test is measured by its ability to detect departures from the null hypothesis. Therefore, in addition to knowing the random sampling distribution of a given test statistic when the null hypothesis is true, it is also desirable to know the distribution of the same statistic when the null hypothesis is false, and some particular alternative hypothesis is true. Using this latter distribution, one may determine the *power* of the test. In the case of the chi-square test, the evaluation of this power involves the use of the *noncentral* chi-square distribution.

5.3.1 The noncentral chi-square distribution

Consider the situation which arises when we test the null hypothesis that the expected values in each frequency class are $NP_0(r)$, when in fact they are $NP_1(r)$, $r = 0, 1, 2, ..., W$. Neyman (1949) has shown that if one holds $NP_0(r)$, $NP_1(r)$, and the significance level fixed, then the power of the test tends to 1 as N increases indefinitely. In order to study the more interesting case in which the power is not close to 1 in large samples, let us adopt Cochran's (1952) argument which assumes that, as $N \to \infty$, the alternative hypothesis approaches the null hypothesis. That is, for each r, let

$$P_1(r) - P_0(r) = \frac{c_r}{(N)^{1/2}},$$

$$NP_1(r) - NP_0(r) = c_r(N)^{1/2}, \tag{5.4}$$

where the c_r remain constant as N increases. Then we may write

$$X_r = \frac{f_r - NP_0(r)}{[NP_0(r)]^{1/2}} = \frac{[f_r - NP_1(r)]}{[NP_1(r)]^{1/2}} \left[\frac{NP_1(r)}{NP_0(r)}\right]^{1/2} + \frac{NP_1(r) - NP_0(r)}{[NP_0(r)]^{1/2}}. \tag{5.5}$$

But

$$\left[\frac{NP_1(r)}{NP_0(r)}\right]^{1/2} = \left[1 + \frac{NP_1(r) - NP_0(r)}{NP_0(r)}\right]^{1/2} = \left[1 + \frac{c_r}{P_0(r)N^{1/2}}\right]^{1/2},$$

and this quantity tends to unity as N becomes large, since c_r and $P_0(r)$ are assumed to be constants. Simultaneously the quantities

$$\frac{f_r - NP_1(r)}{[NP_1(r)]^{\frac{1}{2}}},$$

and therefore the quantities

$$x_r = \frac{[f_r - NP_1(r)]}{[NP_1(r)]^{\frac{1}{2}}} \left[\frac{NP_1(r)}{NP_0(r)}\right]^{\frac{1}{2}},$$

tend to the normal distribution with zero mean and unit variance. At the same time the quantities

$$a_r = \frac{NP_1(r) - NP_0(r)}{[NP_0(r)]^{\frac{1}{2}}}$$

tend to the normal distribution with mean a_r and unit variance. Consequently,

$$X^2 = \sum_{r=0}^{W} \frac{[f_r - NP_0(r)]^2}{NP_0(r)} = \sum_{r=0}^{W} \left\{ x_r + \frac{[NP_1(r) - NP_0(r)]}{[NP_0(r)]^{\frac{1}{2}}} \right\}^2 = \sum_{r=0}^{W} (x_r + a_r)^2,$$

where the x_r are subject to the single linear homogeneous constraint $\sum f_r = N$, or, equivalently,

$$\sum_{r=0}^{W} x_r [NP_1(r)]^{\frac{1}{2}} = \sum_{r=0}^{W} [f_r - NP_1(r)] = 0.$$

Thus, as $N \to \infty$, X^2 is distributed as the sum of squares of independent and approximately normally distributed variables $x_r + a_r$, where the a_r are not all zero. This distribution is known as a *noncentral chi-square distribution*. It is a function of two parameters: the number of degrees of freedom, and a noncentrality parameter $A = (a_0^2 + a_1^2 + ... + a_W^2)$ which has the value

$$A = \sum_{r=0}^{W} \frac{[NP_1(r) - NP_0(r)]^2}{NP_0(r)} = \sum_{r=0}^{W} \frac{c_r^2 N}{NP_0(r)} = \sum_{r=0}^{W} \frac{c_r^2}{P_0(r)}. \tag{5.6}$$

For a simple null hypothesis the variates are subject to a single linear constraint, and the number of degrees of freedom in this case is W. When the $P_0(r)$ have to be estimated from the data, however, the degrees of freedom are further reduced by the number of unspecified parameters.

Tables of the noncentral chi-square distribution have been published by Fix (1949) and Patnaik (1949).

5.3.2 The power function of the chi-square test

Assume that $x_1, x_2, ..., x_N$ are N independent observations drawn from a population which has an unknown distribution function $F(x)$. Suppose we wish to use the chi-square test to test the null hypothesis

$$H_0: F(x) = F_0(x).$$

In order to gain some insight into the power of this test, it is necessary to specify an alternative hypothesis, say

$H_1: F(x) = F_1(x)$.

If H_0 is true, the test statistic X^2 in equation (5.2) will exceed χ_α^2, the α significance point of the chi-square distribution, in a proportion α of the cases. If, however, H_0 is false and H_1 is true, the test statistic will approximately follow the noncentral chi-square distribution, with noncentrality parameter A.

The power of the chi-square test is the probability that X^2 exceeds χ_α^2 under H_1. Denoting this probability by $P_{d.f.}(\chi^{*2}|A)$, the power function is given by

$$\int_{\chi_\alpha^2}^{\infty} P_{d.f.}(\chi^{*2}|A)\,d\chi^{*2} \ .$$

The power, for a given number of degrees of freedom, d.f. say, and significance level α, is a monotonically increasing function of the noncentrality parameter A. Therefore we may express the null hypothesis as

$H_0: A = 0$,

against the alternative hypothesis

$H_1: A > 0$,

where H_1 is a composite hypothesis that includes the family of alternatives for which $\sum_{r=0}^{W} a_r^2 = A$.

The power function can be used in a variety of ways. For example, it can answer questions of the following kind: (a) for a given sample size N, and number of frequency classes $W+1$, what is the probability of correctly rejecting the null hypothesis at the 5% level? (b) For a given number of frequency classes $W+1$, how many observations are required to assure a probability of say 0·90 of establishing significance at the 5% level? Patnaik (1949) illustrates these and other applications of the power function. In chapter 7 we shall use it to define an 'optimal' quadrat size.

Let us illustrate the computation of the power function with an example drawn from quadrat analysis. In such analyses it is frequently the case that spatial distributions, which are well fitted by the Neyman Type A distribution, are also satisfactorily accounted for by the negative binomial. Thus we often end up dealing with the null hypothesis that the empirical distribution is a Neyman Type A distribution against the alternative hypothesis that it is a negative binomial. We have then

$H_0: F(x) =$ **Neyman Type** $A\ (v, a)$,
$H_1: F(x) =$ **Negative binomial** (p, k) . (5.7)

We have seen that the power function depends on three parameters d.f., A, and α. Therefore, for a given sample size, say $N = N^*$, and a prespecified significance level, say $\alpha = \alpha^*$, we can express power as a function of A. However, in the light of hypotheses (5.7) and in recognition of our concern with spatial analysis, a more useful formulation is one which expresses power in terms of the number of points in the point pattern, M say, and the mean and variance of a distribution, or its mean and variance-mean ratio[13]. The latter measure is particularly relevant in that it provides an indication of the degree of clustering which is present in the empirical distribution being analyzed. For a fixed value of M, we may replace A with the mean and variance-mean ratio by computing a value of A for each mean and variance-mean pair. This value of A is then used to associate with each particular mean and variance-mean pair the approximate power of the chi-square test to discriminate between the two hypotheses (5.7). We proceed as follows:

1 Given the population mean $m_1 = m_1^*$ and variance $m_2 = m_2^*$, the variance-mean ratio m_2^*/m_1^* is computed.
2 The expected frequency distribution of the Neyman Type A distribution, for $m_1 = m_1^*$, $m_2 = m_2^*$, and $N = M^*/m_1^*$, is computed.
3 The expected frequency distribution of the negative binomial distribution, for the same values of m_1, m_2, and N, is computed.
4 The noncentrality parameter $A = A^*$ is computed using the expression presented in equation (5.6), and the degrees of freedom are determined by subtracting 1 from the number of frequency classes, $W+1$.
5 A table of the noncentral chi-square distribution, such as Patnaik's, is used to find the power associated with $\alpha = \alpha^*$, $A = A^*$, and d.f. $= W$.

Carrying out these calculations with a numerical example in which m_1, in sequence, takes on the values of $0 \cdot 111$, $0 \cdot 444$, $1 \cdot 000$, $1 \cdot 778$, $2 \cdot 778$, and $4 \cdot 000$, while at the same time m_2/m_1, in the same sequence, takes on the values of $1 \cdot 5$, $2 \cdot 0$, $2 \cdot 5$, $3 \cdot 0$, $5 \cdot 0$, and $10 \cdot 0$, we obtain table 5.1 and its graphical counterpart, figure 5.1[14].

Figure 5.1 illustrates the following very important property of the chi-square test. For a given value of the mean, the power of the test to discriminate between a Neyman Type A distribution and a negative binomial distribution increases with the variance-mean ratio. Since the negative binomial is a more clustered distribution than the Neyman Type A on our dispersion line, it is intuitively apparent that the more clustered the empirical point pattern is, the easier it will be to reject the null

---

[13] Note that a given pair of values for m_1 and M imply, in our problem context, the number and size of the quadrats that cover the study area, for $m_1 = M/N$, whence $N = M/m_1$.

[14] Our choice of values taken on by m_1 reflects an overlay of a grid of $h \times h = N$ quadrats over a study area containing M points. For example, 400 points covered by a 10 by 10 grid of quadrats yields $m_1 = 400/100 = 4 \cdot 0000$, and the same point pattern covered by a 20 by 20 grid leads to $m_1 = 400/400 = 1 \cdot 0000$.

Hypothesis testing: the chi-square test

Table 5.1. Approximate power function of the chi-square test for $M = 400$, $N = M/m_1$, and $\alpha = 0.05$; Neyman Type A versus negative binomial.

m_1	m_2/m_1					
	1·5	2·0	2·5	3·0	5·0	10·0
0·111	0·5	1·0	1·0	1·0	1·0	1·0
0·444	0·2	0·9	1·0	1·0	1·0	1·0
1·000	0·1	0·5	1·0	1·0	1·0	1·0
1·778	0·0	0·2	0·7	0·9	1·0	1·0
2·778	0·0	0·1	0·3	0·8	1·0	1·0
4·000	0·0	0·0	0·1	0·3	1·0	1·0

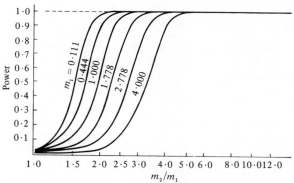

Figure 5.1. Approximate power function of the chi-square test for $M = 400$, $N = M/m_1$, and $\alpha = 0.05$; Neyman Type A versus negative binomial.

hypothesis in (5.7) in favor of the alternative hypothesis. Therefore increasing the variance while holding the mean constant increases power.

5.3.3 The simulated power function

In discussions of hypothesis testing, statisticians commonly distinguish between two kinds of errors: type I errors and type II errors. A type I error is committed when the null hypothesis is rejected, though it is true. A type II error is made when the null hypothesis is not rejected, though it is false. If the distribution of the test statistic X^2 is known under the null hypothesis, a critical region can be chosen for a desired probability of a type I error, say $\alpha = p(I)$. Similarly, if the distribution of the test statistic is known under the alternative hypothesis, one can determine the probability of committing a type II error, say $p(II)$. [Note that power, being the probability of avoiding a type II error, is equal to $1 - p(II)$.]

We have seen that the exact distribution of the test statistic X^2 under H_0 or H_1 is not known. The central and noncentral chi-square distributions, respectively, only provide approximations to its true sampling distribution. Therefore the true probability of making a type I error, p(I), will in general differ from $p^*(I)$, the probability indicated by tables of the χ^2 integral. Similarly the true probability of committing a type II error, $p(II)$, will generally differ from $p^*(II)$, the probability determined by using tables of the noncentral χ^2 integral. In the following paragraphs we report the results of a computer simulation which was developed to test the quality of approximation that is provided by $p^*(I)$ and $p^*(II)$. We shall denote our estimates by $\hat{p}(I)$ and $\hat{p}(II)$ respectively, and, in order to maintain continuity with our earlier discussion, shall focus on $1 - \hat{p}(II)$ (that is, on power) instead of on $\hat{p}(II)$.

We proceeded as follows. First we generated a large number of independent realizations of a distribution known to have a negative binomial distribution with given values of M, m_1, and m_2. We then counted the number of times we rejected the null hypothesis

$H_0: F(x) =$ negative binomial (p, k).

The proportion of times we rejected this null hypothesis was our estimate of $p(I)$. Next, we counted the number of times we rejected the null hypothesis

$H_0: F(x) =$ Neyman Type A (v, a).

The proportion of times we rejected this null hypothesis was our estimate of $1 - p(II)$. We then repeated the entire process for different values of $M, m_1,$ and m_2.

The probability generating function of a negative binomial distribution is

$G(s) = (1 + p - ps)^{-k}$.

The mean of this distribution is $m_1 = kp$, and the variance–mean ratio is $m_2/m_1 = 1 + p$. Given m_1 and m_2, we can find

$$p = \frac{m_2}{m_1} - 1 \quad \text{and} \quad k = \frac{m_1}{m_2/m_1 - 1}.$$

The probability of finding r points in a quadrat is

$$P(r) = \frac{k(k+1)\ldots(k+r-1)}{r!} p^r (1+p)^{-k-r}.$$

For any value of m_1 and m_2, we can compute the cumulative distribution function and, by means of a random number selected between 0 and 1, obtain a realization of the number of points in a quadrat. Repeating the process for N quadrats, and counting the number of cells with 0 points, 1 point, and so on, we arrive at an observed frequency distribution with mean m_1 and variance m_2.

We chose $N = 3600$ (which can be visualized as a 60 by 60 grid of squares). Since in our simulation the number of points in each quadrat is independent of the number in any other quadrat, we can aggregate the N squares, j by j, and find that the number of points in each of the aggregate squares is a sum of independent random variables. Therefore the new probability generating function becomes

$$G_j(s) = (1+p-ps)^{-jk}.$$

Thus we still have a negative binomial distribution, but now with a mean equal to jm_1. Note that the variance-mean ratio and the total number of points remain unchanged, however,

To obtain our realizations, we set j equal to 4, 9, 16, 25, and 36 respectively. This provided us with six realizations for every generated pattern, each with a different mean. We began with $M = 400$. Since $N = 3600$, we had the following six values for the mean m_1:

grid size	60 × 60	30 × 30	20 × 20	15 × 15	12 × 12	10 × 10
mean	0·111	0·444	1·000	1·778	2·778	4·000

Next we set $M = 225$, and obtained for m_1 the values:

| 0·063 | 0·250 | 0·563 | 1·000 | 1·563 | 2·250 |

Finally we set $M = 100$, which yielded the following values for m_1:

| 0·028 | 0·111 | 0·250 | 0·444 | 0·694 | 1·000 |

We then repeated the entire process for the following variance-mean ratios:

| 1·5 | 2·0 | 2·5 | 3·0 | 5·0 | 10·0 |

For each of the realizations we fitted a negative binomial and a Neyman Type A distribution using maximum likelihood estimators. The first fit yielded $\hat{p}(I)$, and the second fit produced $1-\hat{p}(II)$.

Results for $\hat{p}(I)$. After fitting the negative binomial distribution, we computed the value of the X^2 statistic and tested it against the critical value of the chi-square distribution for $\alpha = 0·10, 0·05$, and $0·02$. The proportion of times that this led to a rejection of the true null hypothesis was very close to the expected proportions of $0·10, 0·05$, and $0·02$. The values of m_1 and m_2 did not seem to have any significant influence on $\hat{p}(I)$, but as one would expect, the value of M had a slight impact, the probability of a type I error increasing with decreasing M.

Table 5.2 summarizes the results that were obtained for the three values of M and the three values of α.

Table 5.2. Proportion of times the null hypothesis was rejected when it was true: $\hat{p}(I)$.

M	α		
	0·10	0·05	0·02
400	0·099	0·052	0·020
225	0·102	0·052	0·024
100	0·112	0·052	0·022

Results for $1 - \hat{p}(II)$. Following the process described above, we found that the power of the chi-square test, for a given value of α, varied in the same direction as M and m_2/m_1, but increased when m_1 decreased. Intuitively, the variation with M seems reasonable—as the number of points in the pattern increases, so should the discriminatory power of the test. Similarly, as both the negative binomial and the Neyman Type A distributions tend to the Poisson when m_2/m_1 tends to unity, we would expect the power of the test to increase with m_2/m_1, and this is what we observed. Finally, for M fixed, since $m_1 = M/N$, an increase in m_1 corresponds to a decrease in N, and the power decreases as a consequence. Intuitively this again is reasonable.

In order to test the adequacy of our sample of 100 realizations for obtaining satisfactory approximations of power, we repeated the experiment another 100 times. The results indicated that the addition of 100 realizations did not appreciably change the findings. Hence we conclude that the accuracy obtained by a sample size of 100 is acceptable.

The following three tables summarize the results of our simulations. Table 5.3 provides estimates of $1 - \hat{p}(II)$ for $M = 400$ and different values of m_1, m_2/m_1, and α. Tables 5.4 and 5.5 provide similar information for $M = 225$ and $M = 100$ respectively. Figures 5.2, 5.3, and 5.4 respectively are the graphical counterparts of tables 5.3, 5.4, and 5.5, for $\alpha = 0.05$.

Table 5.3. Simulated power function of the chi-square test for $M = 400$: $1 - \hat{p}(II)$.

m_1	α	m_2/m_1					
		1·5	2·0	2·5	3·0	5·0	10·0
0·111	0·10	0·49	0·88	0·97	1·00	1·00	1·00
	0·05	0·33	0·85	0·96	1·00	1·00	1·00
	0·02	0·23	0·76	0·95	0·98	1·00	1·00
0·444	0·10	0·35	0·71	0·89	0·99	1·00	1·00
	0·05	0·19	0·60	0·87	0·98	1·00	1·00
	0·02	0·11	0·42	0·69	0·91	1·00	1·00
1·000	0·10	0·11	0·41	0·69	0·87	0·99	1·00
	0·05	0·02	0·27	0·63	0·76	0·99	1·00
	0·02	0·00	0·17	0·51	0·69	0·98	1·00
1·778	0·10	0·18	0·19	0·46	0·69	0·96	1·00
	0·05	0·08	0·12	0·35	0·49	0·93	0·99
	0·02	0·03	0·06	0·21	0·31	0·89	0·99
2·778	0·10	0·11	0·16	0·26	0·47	0·84	0·99
	0·05	0·08	0·08	0·15	0·28	0·73	0·99
	0·02	0·03	0·03	0·07	0·15	0·64	0·97
4·000	0·10	0·13	0·13	0·14	0·24	0·64	0·93
	0·05	0·06	0·07	0·09	0·09	0·56	0·92
	0·02	0·02	0·02	0·02	0·04	0·40	0·87

Hypothesis testing: the chi-square test

Table 5.4. Simulated power function of the chi-square test for $M = 225$: $1 - \hat{p}(II)$.

m_1	α	m_2/m_1					
		1·5	2·0	2·5	3·0	5·0	10·0
0·063	0·10	0·47	0·70	0·85	0·95	0·99	1·00
	0·05	0·36	0·55	0·78	0·90	0·98	1·00
	0·02	0·27	0·38	0·65	0·81	0·97	1·00
0·250	0·10	0·26	0·54	0·75	0·90	0·99	1·00
	0·05	0·21	0·36	0·67	0·84	0·99	1·00
	0·02	0·11	0·22	0·46	0·69	0·95	0·99
0·563	0·10	0·17	0·38	0·63	0·70	0·96	1·00
	0·05	0·11	0·26	0·55	0·64	0·92	1·00
	0·02	0·06	0·13	0·35	0·56	0·86	1·00
1·000	0·10	0·17	0·22	0·40	0·61	0·92	1·00
	0·05	0·07	0·13	0·31	0·40	0·86	0·99
	0·02	0·01	0·05	0·18	0·27	0·72	0·97
1·563	0·10	0·11	0·13	0·29	0·46	0·80	0·98
	0·05	0·07	0·08	0·21	0·40	0·66	0·95
	0·02	0·04	0·03	0·15	0·21	0·50	0·89
2·250	0·10	0·09	0·14	0·18	0·28	0·69	0·93
	0·05	0·05	0·06	0·08	0·17	0·57	0·88
	0·02	0·02	0·04	0·03	0·08	0·42	0·78

Table 5.5. Simulated power function of the chi-square test for $M = 100$: $1 - \hat{p}(II)$.

m_1	α	m_2/m_1				
		2·0	2·5	3·0	5·0	10·0
0·028	0·10	0·47	0·46	0·72	0·79	0·89
	0·05	0·26	0·33	0·55	0·73	0·82
	0·02	0·19	0·22	0·36	0·61	0·71
0·111	0·10	0·40	0·44	0·59	0·76	0·87
	0·05	0·24	0·28	0·47	0·68	0·78
	0·02	0·14	0·16	0·30	0·55	0·63
0·250	0·10	0·24	0·33	0·53	0·72	0·85
	0·05	0·15	0·22	0·34	0·60	0·72
	0·02	0·10	0·13	0·21	0·49	0·55
0·444	0·10	0·23	0·27	0·45	0·66	0·82
	0·05	0·12	0·19	0·31	0·56	0·70
	0·02	0·05	0·07	0·14	0·47	0·51
0·694	0·10	0·18	0·20	0·36	0·55	0·79
	0·05	0·06	0·10	0·21	0·37	0·63
	0·02	0·05	0·06	0·07	0·24	0·45
1·000	0·10	0·19	0·17	0·29	0·53	0·74
	0·05	0·09	0·12	0·17	0·38	0·62
	0·02	0·05	0·05	0·09	0·24	0·42

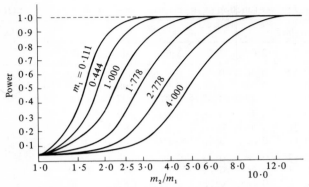

Figure 5.2. Simulated power function of the chi-square test for $M = 400$, $\alpha = 0.05$; Neyman Type A versus negative binomial.

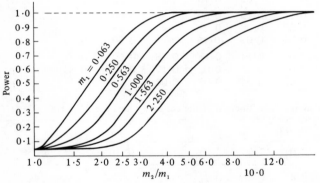

Figure 5.3. Simulated power function of the chi-square test for $M = 225$, $\alpha = 0.05$; Neyman Type A versus negative binomial.

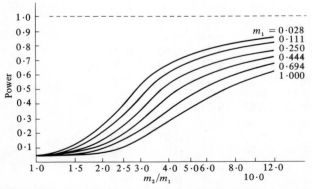

Figure 5.4. Simulated power function of the chi-square test for $M = 100$, $\alpha = 0.05$; Neyman Type A versus negative binomial.

Finally, in figure 5.5, we present, for $m_1 = 1\cdot 000$, a comparison of the power functions for $M = 400$, 225, and 100, as determined by a noncentral chi-square approximation, and as derived by simulation. In figure 5.5, and in figures 5.1 and 5.2, we have clear evidence that the noncentral chi-square approximation always provides an overly generous estimate of power. Moreover the quality of the approximation declines quite markedly with decreasing M, the number of points in the study region.

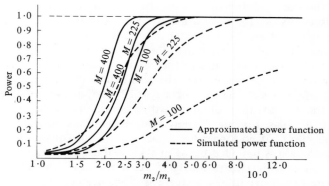

Figure 5.5. Comparison of the approximated and simulated power functions of the chi-square test for $\alpha = 0\cdot 05$, $m_1 = 1\cdot 000$, $M = 400$, 225, and 100; Neyman Type A versus negative binomial.

5.4 Some problems in the use of the chi-square test

Any competent application of the chi-square test must recognize the conditions which restrict the usefulness of the test. One of the more important of these has already been alluded to in our discussion of power. There we noted that chi-square tests are not commonly directed against a specific alternative hypothesis. Hence they cannot point to the way in which the data and the null hypothesis differ, although such differences may be particularly informative. Further, we have seen that the power of the chi-square test to detect a mismatch between theory and data depends on the size of the sample. With a small sample it is very difficult to discriminate between a null hypothesis and an alternative one, because the latter is unlikely to produce a significant value of X^2 in small samples. In a very large sample, however, small departures from the null hypothesis are almost certain to be identified. Thus, in instances of a failure to reject the null hypothesis, our confidence in this hypothesis is directly a function of sample size.

Two other important problems warrant attention. These grow out of the estimation procedure that is used to obtain parameter estimates, and the need to group small expectations prior to the computation of the test statistic.

5.4.1 Estimation problems

Fisher (1924) pointed out that the limiting distribution of the X^2 statistic depends on the method of estimation used to obtain it. An inefficient estimator may lead to a large value of X^2, even if the theory is correct. Fisher also showed that, for large samples, the method which chooses the unknown parameters so as to minimize the value of X^2, becomes equivalent, in the limit, to the method of maximum likelihood. His argument leads to two important results. Any efficient estimator yields estimates which are asymptotically identical to the maximum likelihood estimates; and the χ^2 distribution, with the appropriate reduction in the degrees of freedom, is valid only when efficient estimators are used.

Chernoff and Lehmann (1954) extend Fisher's results. They point out that when the values of the individual observations are available it is more efficient to utilize this knowledge in the parameter estimation process, even though one ends up aggregating the results into the $W+1$ class frequencies before calculating the test statistic. They prove, however, that, when the X^2 statistic is obtained with a maximum likelihood estimation that is based on the N individual observations instead of the $W+1$ frequencies f_r, it no longer has an asymptotic chi-square distribution. Its true distribution is bounded between a χ^2_W and a χ^2_{W-j} variable, where j is the number of estimated parameters. As W increases, the difference between these two distributions decreases sufficiently rapidly for it to be ignored. For small W, however, the use of the χ^2_{W-j} distribution for test purposes may lead to serious error in that the probability of exceeding any given value will be greater than we suppose. Kendall and Stuart (1961) therefore suggest that in such situations the test statistic should exceed the critical values of both distributions.

5.4.2 Grouping problems

In order to prove that X^2 is asymptotically distributed, under the null hypothesis, as the χ^2 distribution, it is necessary to assume that the expectations, $NP_0(r)$, tend to infinity. For this reason, it is customary, in practical applications of the chi-square test, to recommend that the expected number of observations in any frequency class should not be too small. Combination of neighboring classes to satisfy this constraint is therefore generally carried out. However, there is little agreement about the minimum value that should be used. Kendall and Stuart (1961), for example, suggest that all $NP_0(r)$ should not be less than 10. Another lower bound that seems to be popular with statisticians is 5. Cochran (1954), however, argues that in certain situations it is possible to reduce this minimum to 1, without serious loss of accuracy. [Several recent studies seem to support Cochran's argument. For example, see Roscoe and Byars (1971), Slakter (1968), and Yarnold (1970).]

Several difficulties arise as a consequence of grouping. First, the pooling of observations may result in there not being enough frequency classes for a chi-square test to be carried out. That is, there may not be any degrees of freedom left after the reduction for parameter estimates is taken into account. Second, grouping may hide significant deviations between observed and theoretical distributions. For example, grouping may obscure discrepancies from a Poisson distribution and thereby fail to identify nonrandomness when it exists. This may be illustrated by considering once again the numerical example of chapter 4 (see table 5.6).

The observed frequency distribution in this example is a clustered one, as is evidenced by its variance-mean ratio which is greater than 1, and its X^2 statistic, with two degrees of freedom [$NP_0(r) \geq 1$], which is significant at the 5% level ($X^2 = 11 \cdot 40$). Pooling the last 4 frequency classes to maintain a minimum expectation bound of 5, we obtain table 5.7. As a consequence of the above grouping, the X^2 statistic is now equal to $3 \cdot 00$, with 1 degree of freedom. Thus we fail to reject the hypothesis of randomness at the 5% level of significance.

We may generalize the above observations by concluding that grouping reduces power. Loss of power occurs because grouping is generally carried out on classes at the tails or extremes of a distribution. These are frequently the very places where the data deviate most clearly from the hypothetical frequencies. Information about the extent of the loss of power is difficult to obtain, because the power function of the chi-square

Table 5.6. Observed and expected frequencies under the Poisson null hypothesis, $NP_0(r) \geq 1$.

Number of points per quadrat	Observed number of quadrats	Expected number of quadrats under H_0: $F(x)$ = Poisson ($m_1 = 0 \cdot 5200$)
0	67	59·45
1	23	30·92
2	5	8·04
3	2	1·39 ⎫
4	2	0·18 ⎬
5+	1	0·02 ⎭

Table 5.7. Observed and expected frequencies under the Poisson null hypothesis, $NP_0(r) \geq 5$.

Number of points per quadrat	Observed number of quadrats	Expected number of quadrats under H_0: $F(x)$ = Poisson ($m_1 = 0 \cdot 5200$)
0	67	59·45
1	23	30·92
2+	10	9·63

test is known only as a limiting result—a result which assumes that the expectations are large. Nevertheless an approximate measure of loss may be obtained by using tables of the noncentral chi-square distribution.

Consider a sample size of $N = 100$ and the null hypothesis that the numerical observed frequency distribution presented above follows a Poisson distribution with a mean $m_1 = 0 \cdot 5200$. Assume that, in fact, the data have been generated by a negative binomial distribution with mean and variance that are known to be $m_1 = 0 \cdot 5200$ and $m_2 = 0 \cdot 9188$ ($m_2/m_1 = 1 \cdot 7669$) respectively. (Since the expected frequencies are assumed to be known, there are $6 - 1 = 5$ degrees of freedom for the ungrouped case.) To make all $NP_0(r) \geqslant 1$ requires the pooling of the last three frequency classes; to make all $NP_0(r) \geqslant 5$, we must group the last four frequency classes; and to make all $NP_0(r) \geqslant 10$, all but the first frequency class must be aggregated together. Thus the degrees of freedom for the three grouped distributions are 3, 2, and 1 respectively. What is the probability of rejecting the null hypothesis of randomness at the 5% level in each case?

In table 5.8 we present the data and computations which lead to determination of the noncentrality parameter A that is associated with each grouping alternative. Note that for the ungrouped case the last frequency class is the largest contributor to A, and therefore to the power of the test. Grouping the last three frequency classes reduces this contribution, and pooling the last four classes reduces it almost to zero. The approximate probabilities of rejecting the null hypothesis, for each alternative, may be determined by using Patnaik's (1949) table of the noncentral chi-square distribution. They are approximately $1 \cdot 0$, $0 \cdot 8$, $0 \cdot 5$, and $0 \cdot 3$, respectively. The loss of power from grouping is evident.

Table 5.8. Observed and expected frequencies under the null and alternative hypotheses.

No. of points per quadrat	Obs. freq.	Expected frequencies		$[NP_1(r) - NP_0(r)]^2/NP_0(r)$			
		$NP_0(r)$ $H_0: F(x) = P.$	$NP_1(r)$ $H_1: F(x) = N.B.$	Ungrouped	$NP_0(r) \geqslant 1$	$NP_0(r) \geqslant 5$	$NP_0(r) \geqslant 10$
0	67	59·45	67·48	1·08	1·08	1·08	1·08
1	23	30·92	20·55	3·48	3·48	3·48	
2	5	8·04	7·39	0·05	0·05		
3	2	1·39	2·79	1·41		5·59	0·56	1·59
4	2	0·18	1·08	4·50	5·59		
5+	1	0·02	0·70	23·12			
N	100	100·00	100·00	$A = 33 \cdot 64$	$A = 10 \cdot 20$	$A = 5 \cdot 12$	$A = 2 \cdot 67$

The structure of retail trade

6.1 Introduction
The principal activity that takes place in the retail trade sector of an urban economy may be characterized by the input of labor and capital (principally in the form of space) that is used to distribute goods to their ultimate consumers—households. This commercial activity is organized in *establishments*, each with its own productive structure, and the total retail trade sector consists of a population of such establishments. These differ with regard to location, size, and productivity. The aggregate quantities of goods that are transacted in the urban retail sector will depend on the particular structure of the population of retail establishments, their spatial distribution, and the wealth and spatial distribution of their customers.

The retail trade sector of the urban economy has long been a neglected area in urban research, and only recently has it received the scholarly attention it justly deserves. While this is understandable in the light of the historical shift from agricultural to industrial economies (a time during which retailing was of small importance), the emergence of the modern retail and service economy in developed countries has reordered the priorities for economic research. This reordering is manifested in the growth of pioneering studies such as those of Barger (1955), Stigler (1956), Hall *et al.* (1961), and Fuchs (1968).

Although the above mentioned studies have made significant contributions to the body of knowledge concerned with the structure and productivity of retail trade, they are all *nonspatial* in character and approach. That is, they do not consider the pervasive influence of *location* on the structure and productivity of retail trade. Here again the primacy of agriculture and industry is evident. Both agricultural and industrial location theories have a rich history which goes back to the significant contributions of von Thünen (1875) and Weber (1929). Retail location theory on the other hand has been a stepchild of economic research, and its major contributors have come from marketing and geography. Consequently the economics of retail spatial geography still remains to be fully developed.

In this and the next chapter we focus on the statistical measurement of the structure—nonspatial and spatial respectively—of retail trade in urban areas. Although most of our data describe the commercial structure of only a single city (Ljubljana, Yugoslavia), our methods and findings have considerable generality and reveal properties of retail spatial structure that are common to most urban areas in the world.

6.2 Retailing
There are approximately 1 760 000 retail establishments in the United States today (US Census, 1971, p.723). These vary from the small firm offering

small assortments of very limited lines of goods, such as tobacco shops and newsstands, to the huge chain department stores carrying a wide variety of merchandise. Together they accounted for $364 571 000 000 in total retail sales in 1970 (US Census, 1971, p.725). The most important types of retail outlets in terms of retail sales have been the food and automotive groups. In 1970 these together accounted for 40% of all sales. Department stores contributed 11%, and apparel and accessory stores brought in another 6% (US Census, 1971, p.725).

Retail establishments in the United States are typically very small enterprises by any measure of size. In 1967 33% of all stores had sales of less than $30 000 for the year, and these transacted only 3% of all retail business. Another 31% fell into the $30 000-$99 999 class and accounted for 10% of the sales. Thus 64% of all retail stores accounted for only 13% of all sales. In contrast the top 6% category of all stores (that is, stores with sales of $500 000 or more) did 55% of all business (US Census, 1971, p.727). The small size of the average retail outlet becomes even more apparent when the number of employees is used as a measure of size. In 1967 65% of all retail establishments each employed less than 4 employees (US Census, 1971, p.727). Thus it is quite evident that the structure of retail trade is characterized by a large number of very small stores and a very small number of extremely large stores.

6.2.1 Classes of retail goods

Marketing specialists distinguish between three classes of consumer goods: convenience goods, shopping goods, and specialty goods. The standard definitions of these terms reflect the buying habits of the consumer and typically distinguish goods on the basis of the frequency of their purchase and the role brand names play in the purchasing process.

Convenience goods are consumer goods which are bought frequently, and which the consumer therefore desires to purchase with a minimum of effort. Brand names, rather than price, are the principal consideration. Examples are groceries, tobacco, and drug items. Since these goods are purchased day after day, convenience and accessibility are of fundamental importance. Thus stores carrying these goods tend to locate close to the consumer.

Shopping goods are generally acquired at periodic intervals—seasonally, annually, and, in some instances, only once in a lifetime. Such items commonly are unstandardized and are purchased only after considerable comparisons of quality and price have been made in a number of competing retail outlets. Furniture, shoes, jewelry, and style goods in general are prominent examples of this category of consumer goods. To satisfy the consumer's desire to 'shop around', shopping goods stores tend to cluster in major retail centers.

Specialty goods are goods for whose purchase the consumer is willing to make a special purchasing effort. Price once again is not the major consideration, but plays a subordinate role to merchandise quality. Examples are expensive perfumes, watches of superior quality, and high-grade men's and women's clothing. Since consumers of specialty goods typically make a special visit to purchase them, the locational decisions of stores handling such merchandise are quite free of any restraining conditions, such as are met by convenience goods or shopping goods outlets. Thus specialty goods stores may locate virtually anywhere in the urban area without seriously affecting their volume of sales.

It is quite apparent that a considerable degree of overlap exists in the above classification system. Different consumers may, as a result of their buying habits, treat convenience goods as specialty goods and vice versa. Thus, though most people consider groceries as a convenience good, many 'shop around' for them. On the other hand, common work shoes may be treated as a convenience good and purchased at the most convenient location. These difficulties of overlapping categories are inherent in any classification system and once recognized should not seriously hamper most analytical investigations.

6.2.2 Competition

The theory of retail trade describes retailing as monopolistic or imperfect competition (Aubert-Krier, 1954; Lewis, 1945). Although many features of the structure of retail markets are included in the generally accepted definitions of perfect competition, the presence of certain 'imperfections' has led writers to place retail activities in the middle ground between perfect competition and monopoly. These imperfections stem from two major sources: (1) spatial and nonspatial monopolistic influences such as favorable locations or prestige clientele; and (2) the desire of retail firms to minimize risk and limit the area of competitive activities, such as price competition, that can be undertaken by other firms.

Retail establishments continually evaluate the impact that their decisions will have on competitors, and generally pay close attention to the policies of other establishments carrying similar lines of merchandise within the same geographical trading area. This is in contrast to 'perfect competition' where the behavior of a firm has no significant influence on market price. Also it is frequently held that, since people tend to shop around for shopping goods, price elasticity of demand is much greater for shopping goods than convenience goods, and as a result shopping goods retailing is competitive. The same point of view suggests that, since consumers attempt to minimize the disutility of convenience goods shopping, and tend to purchase many convenience goods at one place, convenience goods retailing tends to be imperfectly competitive or monopolistic. This is an oversimplified view of the fundamental difference between the market structure of the two branches of retail trade. However, it does appear to have some merit in explaining the spatial concentration of particular classes of retail firms and the decentralization of others.

6.2.3 Retail location

Retail spatial patterns arise directly out of the locational decisions made by the distributors of retail goods. The individual business firm, operating with incomplete information regarding the urban environment, seeks to serve the spatially scattered market by establishing itself at a profitable location. Analyses of the significant influences bearing on its locational decision form the body of retail location theory (Berry, 1967; Lösch, 1954).

Virtually all of the theories that attempt to analyze, interpret, and ultimately predict the spatial behavior of retail establishments point to a common set of factors, which together are held to influence the locational decisions made by such firms. These are variables having as their principal referent the consumer, the state of the urban system, or the establishment. Related to the first are factors such as the distribution of population, income, and purchasing power. Connected with the second are considerations pertaining to accessibility, site economics, and the availability of advantageous locations. Finally, falling in the third category are factors which reflect the establishment's accommodation of internal needs to the external setting, such as the affinity for other economic activities and the role of spatial monopoly in competition.

The general rationale that emerges concerning the spatial behavior of retail outlets is that the geographical configuration of retail establishments in urban areas reflects the averaging of the attracting and repelling forces of linkages incident to the operation of such firms. Accessibility to the residentially based market, exposure to vehicular or pedestrian movements, and proximity to competing and complementary establishments are typical examples of some of the fundamental considerations that appear to be present in these locational decisions.

Since the size of a retail establishment is limited by the spatial extent of its market area—the delineation of which depends in part on the establishment's size—we would expect studies of retail structure to focus on the joint and simultaneous variation of size and location. Yet this is not generally the case, as Scott (1970, p.60) points out:

"Ideally any analysis of retail markets and the size of establishments should embrace their spatial interrelationships. Yet a wide gulf exists between the theoretical ideal and the reality of most empirical enquiry. A vast literature exists on consumer movement but very little of this is related to specific types of retail outlet classified by trade, organization, and size. On the other hand there is a not inconsiderable literature on the size of retail establishments whether measured by output, input, or an assumed population threshold; but very little of this embodies empirical data on the spatial patterns of consumer movement."

Inevitably, therefore, much of the empirical literature on retail structure is partial, and segmented in that it treats the question of location apart from that of size and vice versa. For example, the central place literature

(Berry and Pred, 1961) abounds in studies that focus on the spatial analysis of retail markets with reference to only the distribution of establishments, classified by retail trade sector, and population. On the other hand the available studies of the productivity, economies of scale, and size distribution in retail trade (Hall *et al.*, 1961; McClelland, 1958) deal with these concepts without reference to spatial location.

The integration of the size dimension with the spatial dimension is beyond the scope of this monograph. Hence the empirical analysis that follows in this and the next chapter reflects the common practice of separate examination of size concentration and spatial dispersion.

6.3 The structure of retail trade: Ljubljana, Yugoslavia

Ljubljana, Yugoslavia, is the study region for this study. This choice reflects the availability there of the detailed and disaggregated data that are necessary for our analysis.

Figure 6.1. Map of Ljubljana, Yugoslavia.

Ljubljana is the capital of the Slovenian Republic in Yugoslavia. The city is built on the site of the Roman town of Emona, which was destroyed early in the fifth century. It has a population of about 200000 people and, along with Zagreb in Croatia, serves as the economic, cultural, and political center of northern Yugoslavia.

In 1966 Ljubljana's public retail sector consisted of 798 retail establishments, of which 490 were located within the 3000 m by 3000 m *study area* centered around the city's central business district [15].

Table 6.1. The retail trade sector in Ljubljana, Yugoslavia: summary statistics for 1966[a].

	Study area	Outer area	Entire region
Food stores			
Number	201	253	454
Volume of sales (10^3 dinars)	301490	204313	505803
Selling space (m^2)	14931	11995	26926
Floor space (m^2)	24904	21063	45967
Number of employees	1405	924	2329
Floor space/employee	17·72	22·81	19·78
Sales/employee	214·58	221·12	217·18
Sales/space (m^2)	12·10	9·70	11·00
Nonfood stores (except 69 tobacco stores)			
Number	239	36	275
Volume of sales (10^3 dinars)	705566	61014	766580
Selling space (m^2)	32772	3033	35805
Floor space (m^2)	44359	4928	49287
Number of employees	2234	184	2418
Floor space/employee	19·86	26·78	20·38
Sales/employee	315·83	331·60	317·03
Sales/space (m^2)	15·91	12·38	15·76
All categories (except 69 tobacco stores)			
Number	440	289	729
Volume of sales (10^3 dinars)	1007056	265327	1272383
Selling space (m^2)	57703	15028	72731
Floor space (m^2)	69263	25991	95254
Number of employees	3639	1108	4747
Floor space/employee	19·03	23·46	20·07
Sales/employee	276·74	239·46	268·04
Sales/space (m^2)	14·54	10·21	13·36

[a] Source: data collected for the author by Lidija Podbregar-Vasle of the Yugoslavian Marketing Research Institute using unpublished data supplied by the Slovenian Statistical Office.

[15] The 3000 m by 3000 m study area was gridded into 3600 square cells, or *quadrats*, each 50 m on a side, for purposes of spatial analysis.

Table 6.2. The retail trade sector in Ljubljana, Yugoslavia: size and 'productivity' distributions in the study area.

a. Employment

Retail subsector	Employees per establishment						
	<3	<6	<9	<12	<15	<20	20+
Food stores	67	60	41	11	4	12	6
Grocery stores	11	21	22	7	3	10	5
Nonfood stores	48	76	45	28	16	12	14
Apparel stores	11	19	17	11	4	4	2

b. Space

Retail subsector	Floor space per establishment (m^2)					
	<50	<100	<200	<300	<400	400+
Food stores	82	64	32	12	4	7
Grocery stores	11	27	22	11	1	7
Nonfood stores	60	67	53	30	12	17
Apparel stores	19	23	11	12	1	2

c. Sales

Retail subsector	Volume of sales per establishment (10^3 dinars)					
	<1000	<2000	<3000	<4000	<5000	5000+
Food stores	124	49	12	6	3	7
Grocery stores	34	26	8	4	3	4
Nonfood stores	73	68	37	25	12	24
Apparel stores	19	22	8	11	5	3

d. Space-labor ratio

Retail subsector	Space-labor ratio per establishment (m^2 per employee)							
	<10	<20	<30	<40	<50	<60	<70	70+
Food stores	48	92	34	10	6	8	1	2
Grocery stores	13	37	20	3	4	1	1	0
Nonfood stores	41	82	67	25	12	4	2	6
Apparel stores	16	23	22	4	0	1	1	1

e. Labor 'productivity'

Retail subsector	Volume of sales per employee by employment size						
	<3	<6	<9	<12	<15	<20	20+
Food stores	226	234	180	152	219	231	231
Grocery stores	240	174	180	164	204	198	224
Nonfood stores	309	473	291	348	319	236	279
Apparel stores	275	285	285	316	294	223	233

f. Space 'productivity'

Retail subsector	Volume of sales per m^2 by employment size						
	<3	<6	<9	<12	<15	<20	20+
Food stores	9	16	11	14	13	12	12
Grocery stores	8	10	9	14	16	10	11
Nonfood stores	11	19	17	16	14	13	16
Apparel stores	14	12	18	16	30	11	16

(Included in the latter total are 50 tobacco shops and kiosks, which will not be considered here.) The remaining 308 establishments in the Ljubljana study region were located outside of the study area—an area that we shall refer to as the *outer area*. A map of Ljubljana with an outline of the study area appears in figure 6.1.

In table 6.1 we see that, of the study region's 729 establishments, 454 are food stores and, of these, 201 are located in the study area. Thus there are more food stores in the outer area than in the study area. Observe, however, that the reverse pattern exists for nonfood stores. Of the 275 nonfood stores in the study region, all but 36 are located inside the study area. Note also the concentration of the region's sales volume, floor space, and employment in the study area.

Of the 201 food stores in the study area, 79 are grocery stores (the remainder being fruit and vegetable shops, butcher shops, bread and milk shops, and confectionaries). Of the 239 nonfood stores, 68 are apparel stores.

Table 6.2 indicates that nonfood stores, on the average, are larger than food stores. The former tend to have more employees, more floor space, and a greater volume of sales, per establishment or per employee. Most of the stores in both categories of retail trade are small by any measure of size, yet our crude measures of 'productivity' do not appear to suggest that such stores are significantly more productive. Only in the case of nonfood stores does there seem to be some evidence that a store with 3 to 5 employees is the most productive scale of operation.

6.3.1 Productivity and size
If technological laws determine how the efficiency of a retail activity varies with the scale of its operation, then they should manifest themselves in the relationships between productivity and size of retail establishments in urban areas. Specifically it should be possible to observe a general tendency of productivity to vary with size, and to identify a definite size (or a size range) that is optimal, as measured for example by cost per unit of output (volume of sales).

Consider a retail activity that is concerned with the marketing of a product or set of products. Assume that the product and the factors used in its merchandising are homogeneous, divisible, and can be measured quantitatively. Assume further that it is possible to identify a functional relationship between the quantities of the input factors that are used, $x_1, x_2, ..., x_n$ (for example, employment), and the quantity of output, y (for example, volume of sales), that is produced. Call such a relationship a *production function* and denote it as $y = g(x_1, x_2, ..., x_n)$. If this production function is continuous, we can define the *marginal productivity* of factor x_1 as $\partial y/\partial x_1$, and its *average productivity* as y/x_1, both of which, in general, will vary with the size of the output. Moreover we can introduce a set of factor prices, $p_1, p_2, ..., p_n$, compute total costs for any factor combination as $\sum p_i x_i$, and with this obtain *unit costs* $\sum p_i x_i / y$ for various volumes of output. If an establishment's unit costs reach a minimum value for a particular scale of operation, then this is its *optimum size*.

In order to compare the production functions of several categories of retail trade, we shall express the production function of each category in terms of the common basic factors of production: labor (employment) and capital (space). Thus, measuring the output in terms of volume of sales V, we shall adopt the production function

$$V = g(L, S),$$

(where L represents labor and S capital) with which we will compare different classes of establishments.

The empirical application of the production function relationship described above raises several severe problems of measurement. First, labor is not a homogeneous factor, and the abilities of individuals to contribute to the output of a retail establishment vary considerably. Second, capital is not merely space but includes a host of other items, such as rent, plant and machinery, display equipment, and delivery vans. Third, sales volume is not a very satisfactory measure of output (McAnally, 1963). The value of the total goods sold as an output by a retail shop cannot be divorced from the value of the merchandise bought as an input, the work of the retailers involved, and the use of the shop premises and other capital equipment. A more desirable measure would be *value added*, that is, the total volume of sales minus labor and capital costs, and minus the wholesale costs of the goods sold. Despite all of the above comments, however, the available data constrain us to adopt employment as our measure of labor input, total floor space as our measure of capital input, and sales volume as our measure of output.

A particular specification of the Cobb–Douglas production function[16] is

$$V = kL^a S^b,$$

where k, a, and b are constants. Such functions are said to be homogeneous to the $(a+b)$th degree because they possess the following property:

$$g(rL, rS) = r^{a+b} g(L, S).$$

That is,

$$V = k(rL)^a (rS)^b = r^{a+b} kL^a S^b.$$

---

[16] For a discussion of the Cobb–Douglas function consult a graduate level text such as Allen (1967).

For the particular case of $a+b = 1$, the production function is said to exhibit constant returns to scale, in which case the efficiency of retail establishments is independent of size, and both small and large establishments may exist even in the presence of perfect competition. If there are increasing returns to scale and $a+b > 1$, then the larger establishments tend to be more efficient. Conversely, if $a+b < 1$, we have decreasing returns to scale, and smaller establishments tend to be more efficient.

Let us now examine the results of fitting a Cobb–Douglas function to selected categories of retail establishments in Ljubljana. First, for grocery stores, we have that $a = 0 \cdot 93758^{**}$ and $b = 0 \cdot 07051$, with a coefficient of determination r^2 that is equal to $0 \cdot 90339$[17]. In contrast, for apparel stores, we have that $a = 0 \cdot 87778^{**}$ and $b = 0 \cdot 13001^{**}$, with $r^2 = 0 \cdot 81771$. It is interesting to note that in both cases we have approximately constant returns to scale—a finding that tends to explain the absence, in table 6.2, of significant trends relating the variation of productivity with size.

Finally, as an example of increasing returns to scale, we may fit the Cobb–Douglas function to data on drug stores to find $a = 0 \cdot 96256^{**}$, $b = 0 \cdot 40215^{**}$, $a+b = 1 \cdot 36471$, and $r^2 = 0 \cdot 59217$.

6.3.2 Size concentration

If economies of scale exist in retailing, then so must a range of optimum size. Consequently we may expect that a majority of the retail stores will be concentrated within this range, because competition should tend to eliminate inefficient establishments. The size distribution of retail establishments in an urban area should, therefore, indicate the range of optimal sizes for each major category of retail trade.

The size of an establishment can be characterized by several measures, some of which are based on inputs (for example, employment) and others on output (for example, volume of sales). In focusing on size distributions we generally are interested in identifying indices that emphasize the differences between units. Such comparisons are often carried out by means of concentration ratios and Lorenz diagrams (Utton, 1970). Some promising results also have been achieved by studying probability distributions that closely approximate the observed size distributions (Hart and Prais, 1956).

Of the many possible ways of measuring concentration levels for purposes of indicating degrees of monopoly and competition in individual retail sectors, perhaps the most widely used measure is the *concentration ratio*. This index is simply the proportion of each retail sector's employment or sales volume that is accounted for by the largest establishments. In the United States this ratio is generally computed for the largest four enterprises.

---

[17] Exponent values that are asterisked once are significantly different from zero at the 10% level. Those that **have two asterisks are significant** at the 5% level.

Applying this concept of the concentration ratio to Ljubljana's retail sector, we find that one of the most concentrated establishment classes with regard to size is grocery trade. The four largest of the 79 grocery stores in the study area accounted for 45·68% of the total volume of sales in the grocery sector in 1966. This may be contrasted with another member of the food category, milk and bread stores, where the four largest of the 45 establishments in this retail class accounted for only 18·45% of the total volume of sales. Corresponding percentages for other selected categories of retail trade are:

Apparel stores (68 shops)   35·93%
Drug stores (45 shops)   32·79%
Shoe stores (34 shops)   24·67%.

A general criticism of the concentration ratio is that it is not a summary measure based on the entire size distribution of establishments in a retail subsector. Nor does it give the relative sizes of the largest four firms. Two categories of retail trade, for example, may both have a four-establishment sales volume concentration ratio of 60%, but in one case the largest establishment may account for 50% of the total, while in the other the largest establishment may only be responsible for 30% of the total. This limitation has led various analysts to suggest other concentration measures that provide fuller information about the complete size distribution of establishments in a particular retail subsector. A popular measure of this kind is the *Lorenz curve*.

A Lorenz curve describes the cumulative percentage of employment (or some other measure of size, such as sales volume) that is accounted for by various *percentages* of establishments in a given retail subsector. Lorenz curves for Ljubljana's milk and bread, grocery, apparel, shoe, and drug stores are set out in figure 6.2. The Lorenz curve of a retail subsector with equally sized establishments would lie on the diagonal rising at a 45° angle from the origin. This line represents a subsector in which x% of the establishments account for x% of whatever index of size is adopted. In figure 6.2 we use employment in the upper triangular portion of the diagram and volume of sales in the lower triangular part.

The further a Lorenz curve is from the 45° 'line of equal distribution', the greater is the size concentration of establishments in the particular retail subsector. Thus, glancing at figure 6.2, we see that grocery stores are once again shown to be more concentrated than the other categories studied, and that milk and bread shops are the closest to being equally distributed.

The search for summary descriptions of size distributions which could be used to compare concentration levels over time and space has more recently led to the use of probabilistic models that purport to describe the stochastic process that may have generated the observed size distribution.

Over the past two decades, the interests of the economist in market concentration and of the statistician in stochastic processes have been fused in several such studies of the size distribution of firms (Hart and Prais, 1956; Simon and Bonini, 1958; Quandt, 1964). Two stochastic models that have provided promising results have been the Pareto and the lognormal distributions (Quandt, 1966; Aitchison and Brown, 1957).

The Pareto distribution is defined by the cumulative distribution function

$$F(x) = 1 - \left(\frac{x_0}{x}\right)^\alpha \qquad x > x_0, \tag{6.1}$$

where x_0 is the lower bound of the domain of x values and α is a parameter. The lognormal distribution has the form

$$F(x) = \int_0^x \frac{1}{\sigma t (2\pi)^{1/2}} \exp\left\{-\frac{1}{2}\left[\frac{\ln(t-\mu)}{\sigma}\right]^2\right\} dt, \tag{6.2}$$

where μ is the mean of $\ln t$ and σ^2 is its variance. The two distributions may be conveniently studied with the aid of logarithmic charts. For example, if the data on size distribution follow the Pareto law, a plot on double logarithmic paper of the percentage of stores with a volume of sales (or some other such measure of size) in excess of each of a set of

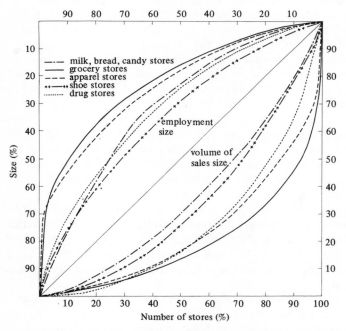

Figure 6.2. Lorenz curves of size concentration for retail establishments in Ljubljana, Yugoslavia.

values against the same values will result in a scatter of points that lie along a negatively sloped straight line with a slope equal to the parameter α in equation (6.1). This can be shown by transposing in equation (6.1) and taking the natural logarithm of both sides:

$$\ln[1 - F(x)] = \alpha \ln x_0 - \alpha \ln x . \tag{6.3}$$

A similar result may be established for the lognormal distribution. If the size distribution data follow a lognormal law, a plot on logarithmic probability paper of the percentage of establishments with a volume of sales, say, in excess of each of a set of values against these same values will result in a scatter of points that lie along a negatively sloped straight line with a slope equal to the reciprocal of the parameter σ in equation (6.2). That is,

$$[1 - F(x)] = \frac{\sigma + \mu}{\sigma} - \frac{1}{\sigma} \ln x . \tag{6.4}$$

Figures 6.3 and 6.4 present the plots of size distribution data on grocery and apparel stores in Ljubljana on double logarithmic paper and on logarithmic probability paper respectively. Since none of the curves is a straight line, we may conclude that the size distributions of the establishments in the selected categories of retail trade do not closely follow either the Pareto or the lognormal distributions. Consequently the stochastic processes that generate these two theoretical distributions do not appear to be satisfactory models of the growth process experienced by the two categories of retail trade.

Both the Pareto and the lognormal distributions have useful properties with respect to their associated Lorenz curves. For example, a line drawn perpendicular to the diagonal line of equal distribution divides the

Figure 6.3. The fit of the Pareto distribution.

lognormal distribution's Lorenz curve into two equal and symmetrical parts. Also it can be shown that the *Lorenz index of concentration, C* say, defined as the ratio of the area between an observed Lorenz curve and the diagonal line of equal distribution to the area of the triangle under the line of equal distribution, depends on α in the Pareto model and on σ in the lognormal model. In the former case,

$$C = \frac{1}{2\alpha - 1}, \tag{6.5}$$

while in the latter case

$$C = 2N\left(\frac{\sigma}{2^{\frac{1}{2}}} \Big| 0, 1\right) - 1, \tag{6.6}$$

where $N(\theta | 0, 1)$ denotes the distribution function of θ, a normally distributed random variable with zero mean and unit variance. Therefore we conclude that smaller values of α (or, equivalently, larger values of σ) indicate higher degrees of concentration in the size distribution. Maximum likelihood estimators of α and σ^2 for our Ljubljana data yield, for example, the following estimates:

	Grocery stores		Apparel stores	
Employees	$\hat{\alpha} = 0\cdot 5578$	$\hat{\sigma}^2 = 0\cdot 8188$	$\hat{\alpha} = 0\cdot 9320$	$\hat{\sigma}^2 = 0\cdot 6213$
Sales	$\hat{\alpha} = 0\cdot 5285$	$\hat{\sigma}^2 = 0\cdot 9104$	$\hat{\alpha} = 0\cdot 5471$	$\hat{\sigma}^2 = 0\cdot 7513$
Space	$\hat{\alpha} = 0\cdot 3643$	$\hat{\sigma}^2 = 1\cdot 0296$	$\hat{\alpha} = 0\cdot 4949$	$\hat{\sigma}^2 = 0\cdot 8642$.

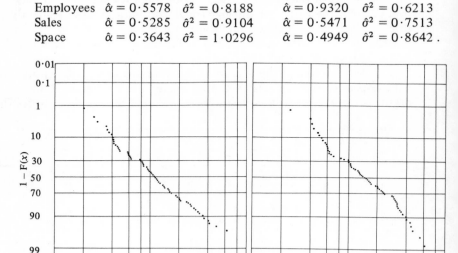

Figure 6.4. The fit of the lognormal distribution.

Thus, once again, we find that grocery stores in Ljubljana exhibit a higher degree of size concentration than apparel stores. This no doubt stems from the tremendous impact wielded by the largest supermarket in downtown Ljubljana, which alone accounts for approximately a fourth of the study area's total volume of sales in the grocery sector. Note also that retail space is more concentrated than retail sales, which in turn are more concentrated than retail employment.

6.3.3 Spatial concentration

The extension of the concepts of size concentration to spatial concentration is a straightforward task methodologically, but a somewhat more complex task conceptually. The existence of spatial regularities in retail patterns is present in most urban areas, but the interpretation of such regularities must be approached with caution.

The spatial location of an establishment can be conveniently characterized by its coordinates on a map. In focusing on the spatial distribution of such points, we are interested, as in the case of size distributions, in measures that identify differences between retail subsectors. Such comparisons once again may be carried out by means of concentration ratios, Lorenz curves, and stochastic models.

Consider the extension of the concentration ratio, described earlier in the discussion of size concentration, to the case of spatial concentration. In place of the *technological unit* of the establishment, let us now focus on the *spatial unit*: a 50 m by 50 m cell, the fundamental grid in the 3000 m by 3000 m study area discussed earlier. Of the 3600 cells in that study area only 259 cells have retail establishments in them. That is, 92·81% of the study area's 3600 cells have not a single retail establishment. Disaggregating this number down into specific categories of retail trade, we have the following numbers of cells that are occupied by at least one establishment in that retail category:

All stores	259 (7·19%)
Food stores	165 (4·58%)
Nonfood stores	143 (3·97%)
Grocery stores	78 (2·17%)
Apparel stores	53 (1·47%) .

The above concentration ratios, however, have the defect that they are dependent on the size of the cell. For example, aggregating the cells into a grid of 400 cells that are 150 m on a side leads to the following different set of results:

All stores	114 (28·50%)
Food stores	96 (24·00%)
Nonfood stores	65 (16·25%)
Grocery stores	59 (14·75%)
Apparel stores	32 (8·00%) .

As in the case of size concentration we may focus on the number of stores contained in the 'largest', or in this instance the densest, cells. For example, of the 400 cells that are 150 m on a side, the four 'densest' cells contain 84 of the 440 retail establishments in the study area (19·09%). The four densest cells with food stores, contain 32 of the 201 food stores (15·92%), those with nonfood stores contain 84 of the 239 nonfood stores (35·15%), those with grocery stores contain 12 of the 79 grocery stores (15·19%), and those with apparel stores account for 29 of the 68 apparel stores (42·65%).

These concentration levels, high as they are, nevertheless are biased toward the low side for they deal with numbers of establishments. Since these establishments are themselves highly concentrated with regard to size, we may wish to approach the problem of spatial concentration with reference to employment or sales volume. For example, returning once more to the 400 cells, 114 of which contain retail shops, we find that the four 'densest' cells account for 38·02% of the total sales volume in the study area and for 56·25% of its employment. Analogous results may be computed for specific categories of retail trade:

	Sales volume	Employment
All stores	38·02%	56·25%
Food stores	37·67%	34·09%
Nonfood stores	38·17%	70·19%

Note that, although the sales volume concentration levels are about the same for food and nonfood stores, the latter's employment concentration level is more than double that of the former.

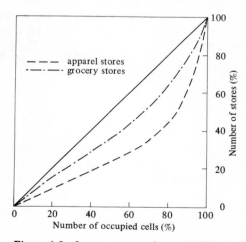

Figure 6.5. Lorenz curves of spatial concentration for retail establishments in Ljubljana, Yugoslavia.

The Lorenz curve also may be used to describe the relative concentration levels of retail trade in space (Rannells, 1956). Focusing on the number of cells that are occupied by at least one retail establishment of a particular category, say apparel stores, we simply compute the cumulative percentage of the total number of such establishments that are accounted for by various percentages of the total number of 'occupied' cells. Thus in the case of apparel stores, for example, we have already seen that the densest four cells (or 1% of the study area) account for 42·65% of the 68 stores. Adding on the fifth cell increases this total to 50·00%, and proceeding in such a manner we may define the Lorenz curve for apparel stores that appears in figure 6.5, and contrast it with the corresponding one for grocery stores. Analogous results may be found for other categories of trade and for other measures of size, such as employment or sales volume.

Paralleling the fusion of the interests of economists and statisticians that produced the currently available stochastic models of size distribution has been the fusion of the interests of geographers, ecologists, and statisticians that has produced the currently available stochastic models of spatial dispersion. These models were outlined in the first half of this book, and their application to our data for Ljubljana, Yugoslavia, is the subject of the next chapter.

The spatial dispersion of retail trade in urban areas

7.1 Introduction

Despite the great physical changes brought about by the tremendous growth of suburban areas, the retail spatial structure of most urban areas has maintained a remarkably constant profile during the past thirty years. Retail establishments in most major cities generally belong to one of the following classes of centers: the central business district, outlying secondary shopping districts, neighborhood business nucleations or small retail clusters, and isolated individual stores.

The tendency for particular classes of retail establishments to cluster and form such centers has been noted by several theorists on the subject of intraurban retail spatial structure. Hotelling (1929), for example, in his classical essay on spatial competition, concludes that as "more and more sellers of the same commodity arise, the tendency is not to become distributed in the socially optimum manner but to cluster unduly". Nelson (1958) supplies a rationale for this tendency with his "principle of cumulative attraction". This principle holds that a group of stores carrying the same merchandise will do more business if they are clustered together than if they are widely scattered. Thus, for example, a nucleus of three shoe stores will do a larger volume of business than will the same three located several blocks apart.

Reference to the generally regular pattern of convenience goods outlets in urban areas also appears in much of the theoretical literature on retail spatial structure. Ratcliff (1939), for example, recognizes that certain neighborhood type establishments, such as grocery stores, do not profit from cohesion and therefore typically do not group together. Others assert that the scattered pattern of convenience goods stems from their limited 'range' and the restricted market areas that they command as a result of this characteristic (Berry and Garrison, 1958a, 1958b). Finally marketing researchers, such as Nelson (1958), note that a convenience goods establishment, in selecting a location, should be primarily concerned with accessibility and apparent freedom from future competition.

It appears then that explanations of the spatial pattern of consumer goods firms typically cite two interdependent factors as the principal contributory agents to clustering or regularity: the nature of consumer goods and the structure of consumer goods markets. Catering to the immediate every day needs of the consumer, convenience goods establishments have followed the urban population to the outlying areas of metropolitan communities and have assumed a spatial pattern which mirrors that of the population they serve. The day after day needs for such goods have made accessibility and proximity of fundamental importance in the purchasing process, and the hazards of competition encourage dispersal. **Shopping** goods establishments, however, frequently reflect the thesis that firms with a high degree of compatibility can ensure a greater volume of sales for themselves by clustering together. Their compatibility may stem from the complementary nature of their activities and products, or may occur when, though competing with each other, they carry goods of different quality and price. Thus, in a very general sense, we may conclude that rival shopping goods establishments tend to attract one another and thus to assume a more clustered spatial pattern. Establishments falling outside this dichotomous classification may be viewed as distributors of specialized goods and may be assumed to have locational characteristics that are appropriate for their particular activity. Typical examples are traffic oriented activities, such as gasoline service stations, which appear to be highly dependent on the traffic flow pattern of an urban area.

7.2 Compound and generalized distributions as models of retail spatial dispersion

Compound and generalized distributions offer a variety of models that can be fitted to empirical data. Both classes of models assume that a given urban area has been gridded into a network of squares, and both account for the observed dispersion in this network by describing the spatial regularity or clustering of retail establishments as a realization of a general stochastic process. Compound distributions postulate areal inhomogeneity of some important parameter such as the mean. Generalized distributions reflect the assumption that stores occur only in clusters.

7.2.1 The compound model

Assume that in an urban area which has been gridded into a network of square quadrats the number of stores in each cell, R_1 say, may be described by the probability distribution $P_1(r_1)$. Next, assume that some parameter of $P_1(r_1)$, R_2 say, varies from place to place, within the study area, according to the probability distribution $P_2(r_2)$. Then $P_1(r_1)$ is a conditional probability distribution that depends on the values assumed by R_2, and therefore can be denoted as $P_1(r_1|r_2)$. Thus $P(r)$, the unconditional probability distribution of the number of stores in each quadrat, may be expressed as

$$P(r) = \int_{-\infty}^{\infty} P_1(r|r_2) P_2(r_2) \mathrm{d}r_2 \,, \tag{7.1}$$

which, except for the missing arbitrary constant c, is the compound distribution defined in equation (3.1).

7.2.2 The generalized model

Given an urban area that has been gridded into a network of square quadrats, consider a model that seeks to describe the spatial dispersion of stores within this network by means of a general stochastic process based on the following postulates:
1 Stores occur only in clusters.

2 The number of clusters in a quadrat, R_1 say, is described by the probability distribution $P_1(r_1)$.
3 The number of stores in a cluster, R_2 say, varies from one cluster to another according to the probability distribution $P_2(r_2)$, which is the same for all clusters.

The statement of the problem of retail spatial dispersion in these terms identifies the process as a generalized distribution with the probability generating function defined in equation (3.7).

7.3 The spatial dispersion of retail establishments in Ljubljana, Yugoslavia

Several sets of data will be used for the empirical testing of the distributions that have been defined in this study. The most detailed data describe the spatial pattern of retail establishments in Ljubljana, Yugoslavia, and we begin our analysis with a careful study of the spatial dispersions exhibited by different subsectors of retail trade in that city.

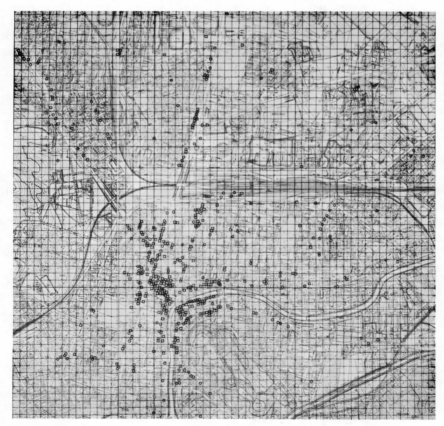

Figure 7.1. The spatial pattern of retail establishments in Ljubljana, Yugoslavia.

7.3.1 Intraurban comparisons

In chapter 6 we saw that 440 of the 720 retail establishments considered in our study region of Ljubljana were located inside the 3000 m by 3000 m study area. Figure 7.1 presents the spatial pattern of these 440 retail establishments; 201 of these were food stores, a category that included 79 grocery stores (the remainder being fruit and vegetable shops, butcher shops, stores selling only bread and milk, confectionaries, etc.). Of the 239 nonfood stores, 68 stores were classified as apparel stores (stores primarily engaged in selling clothing, shoes, hats, etc.). Figure 7.2 presents the spatial patterns of these four categories of retail establishments.

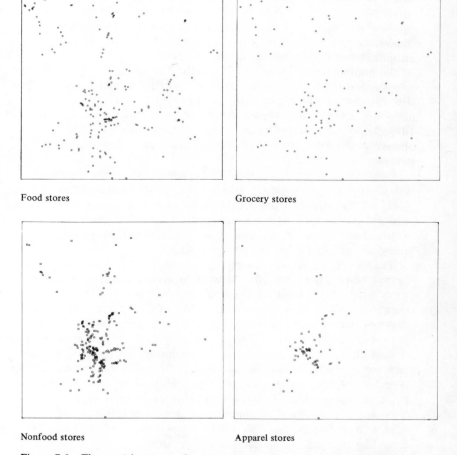

Figure 7.2. The spatial pattern of retail establishments in Ljubljana, Yugoslavia.

The study area may be gridded in a number of different ways for purposes of quadrat analysis. The 3000 m by 3000 m study area can be viewed as a 60 by 60 grid of squares, 50 m on each side, and then these may be aggregated to form 30 by 30, 20 by 20, 15 by 15, 12 by 12, and 10 by 10 grid systems. For each of these observed data sets, parameter estimates may be derived by the method of maximum likelihood, and for each distribution the relevant expected values can be calculated. The adequacy of the fit provided by each of the theoretical frequencies may then be tested by means of the chi-square goodness of fit test. To eliminate the disproportionate effect of small differences from a low expectation, we shall pool frequencies with small expectations such that no expectation is less than 5. And to condense an otherwise unwieldy array of results, we shall only consider the fits provided by one quadrat size—the 250 m by 250 m cell of the 12 by 12 grid system. This was the quadrat size used by Artle (1965) in his study of Stockholm's retail dispersion and will therefore afford us an opportunity for carrying out an interurban comparison of retail dispersion. Table 7.1 presents the fits of the binomial, the Poisson, the negative binomial, the Neyman Type A, the Poisson-binomial, and the Poisson-negative binomial distributions to the data on the spatial dispersion of Ljubljana's stores—all categories, food, nonfood, grocery, and apparel stores—when sampled with a square grid of 144 quadrats 250 m on a side. Table 7.2 provides a complete list of observed and expected frequencies for one class of establishment—food stores.

For each of the five empirical dispersions investigated the negative binomial distribution clearly provides the most satisfactory fit. Of the nine null hypotheses examined in each instance, this one alone is never rejected at the 5% level of significance. Thus one may, with reasonable justification, conclude that this theoretical distribution provides an adequate accounting for the observed data.

The theoretical model generating the negative binomial, however, cannot be clearly identified. As we have seen there are a number of ways in which the negative binomial distribution can arise. One cannot, therefore, on the basis of an observed set of data alone, distinguish between the different probability mechanisms that may be operating to produce this distribution. Thus, for example, it was shown in chapter 3 that, if clusters of establishments are randomly distributed and the number of establishments in each cluster follows a logarithmic distribution, then the resulting number of stores per cell is given by the negative binomial distribution. It was also demonstrated there that the same distribution will arise if the number of stores per cell does follow a random distribution, but one in which the mean is allowed to vary according to a gamma probability distribution.

Table 7.1. Parameter estimates and the goodness of fit for various classes of retail establishments in Ljubljana, Yugoslavia.

a. All categories

Model	$\hat{m}_1 = 3\cdot 0556$	$\hat{m}_2 = 53\cdot 0458$	$\hat{m}_2/\hat{m}_1 = 17\cdot 3605$	X^2	d.f.
Binomial	$\hat{n} = 51$		$\hat{p} = 0\cdot 0559$	$898\cdot 20^*$	5
Poisson	$\hat{v} = 3\cdot 0556$			$831\cdot 15^*$	6
Negative binomial	$\hat{w} = 3\cdot 0556$		$\hat{k} = 0\cdot 2255$	$4\cdot 41$	6
Neyman Type A	$\hat{v} = 0\cdot 7765$		$\hat{a} = 3\cdot 9351$	$46\cdot 27^*$	7
Poisson-binomial ($n = 2$)	$\hat{v} = 1\cdot 7115$		$\hat{p} = 0\cdot 8926$	$154\cdot 46^*$	5
Poisson-binomial ($n = 3$)	$\hat{v} = 1\cdot 2618$		$\hat{p} = 0\cdot 8072$	$69\cdot 74^*$	5
Poisson-binomial ($n = 4$)	$\hat{v} = 1\cdot 0949$		$\hat{p} = 0\cdot 6977$	$59\cdot 18^*$	6
Poisson-binomial ($n = 5$)	$\hat{v} = 0\cdot 9978$		$\hat{p} = 0\cdot 6125$	$47\cdot 90^*$	6
Poisson-negative binomial	$\hat{v}_1 = 1\cdot 0630$	$\hat{k}_1 = 0\cdot 2132$	$\hat{p}_1 = 13\cdot 4859$	$12\cdot 66^*$	4

b. Food stores

Model	$\hat{m}_1 = 1\cdot 3958$	$\hat{m}_2 = 6\cdot 9261$	$\hat{m}_2/\hat{m}_1 = 4\cdot 9620$	X^2	d.f.
Binomial	$\hat{n} = 22$		$\hat{p} = 0\cdot 0634$	$139\cdot 10^*$	3
Poisson	$\hat{v} = 1\cdot 3958$			$123\cdot 09^*$	3
Negative binomial	$\hat{w} = 1\cdot 3958$		$\hat{k} = 0\cdot 3544$	$3\cdot 99$	4
Neyman Type A	$\hat{v} = 0\cdot 6435$		$\hat{a} = 2\cdot 1691$	$4\cdot 14$	4
Poisson-binomial ($n = 2$)	$\hat{v} = 0\cdot 8973$		$\hat{p} = 0\cdot 7778$	$34\cdot 62^*$	3
Poisson-binomial ($n = 3$)	$\hat{v} = 0\cdot 7353$		$\hat{p} = 0\cdot 6328$	$14\cdot 71^*$	3
Poisson-binomial ($n = 4$)	$\hat{v} = 0\cdot 6930$		$\hat{p} = 0\cdot 5035$	$10\cdot 66^*$	3
Poisson-binomial ($n = 5$)	$\hat{v} = 0\cdot 6688$		$\hat{p} = 0\cdot 4174$	$8\cdot 72$	3
Poisson-negative binomial	$\hat{v} = 1\cdot 2807$	$\hat{k} = 0\cdot 4709$	$\hat{p} = 2\cdot 3147$	$1\cdot 66$	3

c. Grocery stores

Model	$\hat{m}_1 = 0\cdot 5486$	$\hat{m}_2 = 1\cdot 0745$	$\hat{m}_2/\hat{m}_1 = 1\cdot 9587$	X^2	d.f.
Binomial	$\hat{n} = 5$		$\hat{p} = 0\cdot 1097$	$8\cdot 64^*$	1
Poisson	$\hat{v} = 0\cdot 5486$			$5\cdot 47^*$	1
Negative binomial	$\hat{w} = 0\cdot 5486$		$\hat{k} = 0\cdot 6374$	$1\cdot 49$	1
Neyman Type A	$\hat{v} = 0\cdot 8975$		$\hat{a} = 0\cdot 6112$	$2\cdot 61$	1
Poisson-binomial ($n = 2$)	$\hat{v} = 0\cdot 6913$		$\hat{p} = 0\cdot 3968$	$4\cdot 18^*$	1
Poisson-binomial ($n = 3$)	$\hat{v} = 0\cdot 7523$		$\hat{p} = 0\cdot 2431$	$3\cdot 19$	1
Poisson-binomial ($n = 4$)	$\hat{v} = 0\cdot 7877$		$\hat{p} = 0\cdot 1741$	$2\cdot 95$	1

Table 7.1 (continued)

c. Grocery stores (continued)

Poisson–binomial ($n = 5$)	$\hat{v} = 0.8084$		$\hat{p} = 0.1357$	2.85	1
Poisson–negative binomial	$\hat{v}_1 = 1.4676$	$\hat{k}_1 = 0.6392$	$\hat{p}_1 = 0.5848$	(7.91^*)	2

d. Nonfood stores

Model	$\hat{m}_1 = 1.6597$	$\hat{m}_2 = 29.7086$	$\hat{m}_2/\hat{m}_1 = 17.8997$	X^2	d.f.
Binomial	$\hat{n} = 43$		$\hat{p} = 0.0386$	304.75^*	3
Poisson	$\hat{v} = 1.6597$			290.26^*	3
Negative binomial	$\hat{w} = 1.6597$		$\hat{k} = 0.1073$	3.49	3
Neyman Type A	$\hat{v} = 0.4205$		$\hat{a} = 3.9475$	39.84^*	5
Poisson–binomial ($n = 2$)	$\hat{v} = 0.9391$		$\hat{p} = 0.8836$	94.19^*	3
Poisson–binomial ($n = 3$)	$\hat{v} = 0.7095$		$\hat{p} = 0.7798$	71.99^*	3
Poisson–binomial ($n = 4$)	$\hat{v} = 0.6096$		$\hat{p} = 0.6806$	40.88^*	3
Poisson–binomial ($n = 5$)	$\hat{v} = 0.5610$		$\hat{p} = 0.5917$	35.62^*	3
Poisson–negative binomial	$\hat{v}_1 = 0.2616$	$\hat{k}_1 = 0.6010$	$\hat{p}_1 = 10.5556$	32.50^*	1

e. Apparel stores

Model	$\hat{m}_1 = 0.4722$	$\hat{m}_2 = 3.8174$	$\hat{m}_2/\hat{m}_1 = 8.0839$	X^2	d.f.
Binomial	$\hat{n} = 16$		$\hat{p} = 0.0295$	29.76^*	1
Poisson	$\hat{v} = 0.4722$			28.53^*	1
Negative binomial	$\hat{w} = 0.4722$		$\hat{k} = 0.1038$	3.53	1
Neyman Type A	$\hat{v} = 0.2426$		$\hat{a} = 1.9464$	12.31^*	2
Poisson–binomial ($n = 2$)	$\hat{v} = 0.3335$		$\hat{p} = 0.7081$	(17.21^*)	2
Poisson–binomial ($n = 3$)	$\hat{v} = 0.2843$		$\hat{p} = 0.5537$	11.78^*	1
Poisson–binomial ($n = 4$)	$\hat{v} = 0.2658$		$\hat{p} = 0.4442$	12.00^*	1
Poisson–binomial ($n = 5$)	$\hat{v} = 0.2598$		$\hat{p} = 0.3635$	11.90^*	1
Poisson–negative binomial	$\hat{v}_1 = 0.2008$	$\hat{k}_1 = 0.4970$	$\hat{p}_1 = 4.7320$	(28.72^*)	5

$\hat{v}_1, \hat{k}_1, \hat{p}_1$ are moment estimates of v, k, and p.
* denotes X^2 statistic significant at 5% level.
(X^2) is the X^2 statistic computed with grouping ≥ 1 instead of ≥ 5.

Table 7.2. Observed and expected distributions of food stores in Ljubljana, Yugoslavia.

Number of stores per cell	Number of cells observed	B.	P.	N.B.	N.T.A.	P.B.(2)	P.B.(3)	P.B.(4)	P.B.(5)	P.N.B.
0	83	34·05	35·66	81·76	81·44	61·37	71·59	75·11	77·16	82·89
1	18	50·74	49·77	23·11	12·99	19·03	13·47	12·83	12·41	19·85
2	13	36·09	34·74	12·48	15·12	36·26	24·49	20·61	18·78	12·57
3	9	16·30	16·16	7·81	12·49	10·64	17·79	16·59	15·65	8·50
4	7	5·25	5·64	5·22	8·55	10·67	6·69	8·42	8·89	5·89
5	7	1·28	1·57	3·63	5·37	2·97	5·30	4·64	4·83	4·13
6	2	0·25	0·07	2·58	3·29	2·08	2·55	2·93	2·93	2·92
7	1	0·04	0·01	1·87	1·99	0·55	1·09	1·47	1·66	2·07
8	2			1·37	1·18	0·30	0·60	0·72	0·85	1·48
9	0			1·01	0·69	0·08	0·25	0·36	0·43	1·05
10+	2			3·16	0·89	0·05	0·18	0·32	0·41	2·63
		$X^2 =$ 139·10	123·09	3·99	4·14	34·62	14·71	10·66	8·72	1·66
		$P_{0·05} =$ 7·82	7·82	9·49	9·49	7·82	7·82	7·82	7·82	7·82

Total number of cells = 144
Total number of stores = 201

It is highly probable that both mechanisms are operative in the retail activity system. Retailers carrying the same class of shopping goods merchandise do indeed appear to be attracted to one another; however, they also are attracted to purchasing power. Since purchasing power is unevenly distributed across the urban landscape, one may with some justification assert that the number of retail establishments that locate in a particular cell is distributed according to a Poisson distribution, but not necessarily with the same mean. That is, it could be held that the average number of stores per cell varies according to the purchasing power present in that cell. Thus, if purchasing power is distributed according to a gamma distribution, the negative binomial distribution describes the number of stores per cell.

We recall that the negative binomial distribution is one of a series of distributions that may be used to account for data in which the variance is larger than the mean. The variance of the negative binomial is

$$m_2 = m_1 + \frac{m_1^2}{k},$$

where m_1 is the mean and k the exponent. Hence it is clear that, as $k \to \infty$ while m_1 remains fixed, the variance m_2 approaches the mean m_1. That is, as k increases, the ratio of the variance over the mean approaches unity. Indeed in chapter 2 we showed that in such instances the distribution itself converges to the Poisson distribution. On the other hand, as $k \to 0$ from the positive direction the variance increases without bound and greatly exceeds the mean indicating a highly clustered distribution.

It appears therefore that the relative value of the parameter k in the negative binomial distribution may be used to indicate the 'spatial affinity' that firms in a particular class of retail establishments have for one another. For the Ljubljana data, the following estimates for k were derived:

Apparel stores	$\hat{k} = 0\cdot1038$
Nonfood stores	$\hat{k} = 0\cdot1073$
All stores	$\hat{k} = 0\cdot2255$
Food stores	$\hat{k} = 0\cdot3544$
Grocery stores	$\hat{k} = 0\cdot6374$.

Thus apparel stores seem to exhibit the highest degree of clustering and grocery stores the least. Almost identical findings are indicated by the corresponding nearest-neighbor statistics, $\bar{d}$ and D^*, defined in chapter 1:

Apparel stores	$\bar{d} = 0\cdot80$	$D^* = 0\cdot4393$
Nonfood stores	$\bar{d} = 0\cdot37$	$D^* = 0\cdot3842$
All stores	$\bar{d} = 0\cdot32$	$D^* = 0\cdot4547$
Food stores	$\bar{d} = 0\cdot54$	$D^* = 0\cdot5091$
Grocery stores	$\bar{d} = 1\cdot46$	$D^* = 0\cdot8674$.

We may locate the five classes of retail establishments on the dispersion line, as in figure 7.3.

Note that the ordering of clothing and nonfood stores defined by the parameter k is reversed by the nearest-neighbor index D^*. Are they in fact measuring different attributes, or is the apparent contradiction a consequence of a change in some important factor? It appears that both indices measure approximately the same attribute, but whereas the former index is independent of the density of establishments in a study area, the latter index is not and reflects an increase in 'crowding'. This may be demonstrated by equalizing the number of establishments in the two categories of retail trade, and then observing the responses of the two indices to this change. Deleting 171 randomly selected nonfood stores from the 239 nonfood stores that are in our study area would leave the parameter k unchanged, but it would decrease the value of D^* (Pielou, 1969, pp.90–98).

Comparisons of the spatial clustering of two or more classes of retail establishments usually involve populations that differ in their mean density as well as in their degree of clustering. The advantage of using k as an index of clustering in such instances is that it identifies an intrinsic property of spatial dispersion whatever the density of the population being studied[18].

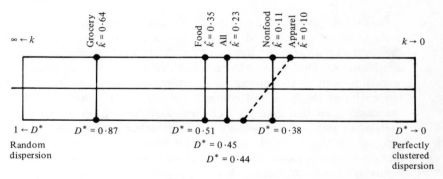

Figure 7.3. Location of the five classes of retail establishments in Ljubljana, Yugoslavia, on the dispersion line.

---

[18] Professor J. B. Douglas, in a private communication, has suggested the use of the ratio $C = m_1/k$ as a more appropriate index of clustering, since a large C implies higher clustering whether or not m_1 is constant. Using this index, one obtains the following values for C: apparel, 4·6; nonfood, 15·5; total, 13·5; food, 3·9; grocery, 0·9. This implies that total and nonfood stores are more highly clustered than apparel stores. It appears that, roughly speaking, C reflects the mean number of stores per cluster, and k the number of clusters per quadrat. Thus his index of clustering favors a 'many-stores-per-cluster' interpretation of spatial clustering, whilst ours advances a 'few-clusters-per-quadrat' interpretation.

The clustered spatial dispersion exhibited by food stores and grocery stores at first appears to contradict the common hypothesis in studies of retail structure that convenience goods stores tend to repel one another. However, it is generally accepted that convenience goods stores are strongly influenced in their locational decisions by the spatial distribution of population in an urban area. Thus, if this influencing factor were spatially clustered, its effect on the locational pattern of convenience goods stores would be to cluster them in spite of their hypothesized repelling effect on one another. That is, the apparent spatial clustering of convenience goods outlets may be due to the clustered distribution of population in urban areas and not the result of a spatial affinity between rival establishments. To investigate this possibility, a crude means of measuring the spatial clustering of population in the study area was devised. The data on population distribution were recorded using class intervals with a 'width' of 500 persons each. Thus the logical transfer of these data into a spatial pattern was to consider every point within the pattern as representing 500 people. This is very crude, to be sure, but it is the only way of utilizing the available information. The resulting spatial dispersion of the residential population in the Ljubljana study area was then fitted by the same methods as were all the other observed frequency distributions of retail establishments. The results are summarized in tables 7.3 and 7.4.

On the basis of this crude analysis it appears that the residential population in the study area is clustered. This is suggested by the close fit provided by the negative binomial, the Neyman Type A and the Poisson–binomial distributions. Thus there exists considerable justification for retaining the original hypothesis concerning the agglomerative and disagglomerative tendencies of shopping and convenience goods establishments. However, the implication that emerges from the above discussion is that any analysis of the mutual repulsion of convenience goods outlets must separate out the effect of population attraction. In other words it appears that convenience goods retailers are under the influence of at least two significant locational forces. Because of the unspecialized and recurrent function they perform, such producers tend to be drawn toward the consumers they serve and thus to assume a distributional pattern similar to that of the residential population. At the same time, however, the "locational relation among producers competing for markets is generally one of mutual repulsion represented by the efforts of each seller to find a market where there is not too much competition" (Hoover, 1948, p.48). Thus, to the extent that these locational considerations conflict—as they do in instances of a spatially clustered population—convenience establishments will tend to distribute themselves in an equilibrium pattern in which the conflicting locational considerations are averaged out with the more dominant consideration counteracting and outweighing the effects of the other. Hoover (1948, p.48, note 2) **suggests** a particularly

Table 7.3. The spatial dispersion of the residential population in Ljubljana, Yugoslavia: parameter estimates and the goodness of fit.

Size of population per square (number of persons)	Number of '500-person' units per square	Number of squares with the indicated size of residential population
0–500	0	90
501–1000	1	28
1001–1500	2	15
1501–2000	3	9
2001–2500	4+	2

Model	$\hat{m}_1 = 0.6458$	$\hat{m}_2 = 0.9856$	$\hat{m}_2/\hat{m}_1 = 1.5261$	X^2	d.f.
Binomial	$\hat{n} = 4$		$\hat{p} = 0.1615$	21.69[a]	1
Poisson	$\hat{v} = 0.6458$			13.60[a]	1
Negative binomial	$\hat{w} = 0.6458$		$\hat{k} = 0.9118$	1.59	1
Neyman Type A	$\hat{v} = 0.9521$		$\hat{a} = 0.6783$	0.31	1
Poisson–binomial ($n = 2$)	$\hat{v} = 0.6882$		$\hat{p} = 0.4692$	2.40	1
Poisson–binomial ($n = 3$)	$\hat{v} = 0.7134$		$\hat{p} = 0.3018$	0.61	1
Poisson–binomial ($n = 4$)	$\hat{v} = 0.7614$		$\hat{p} = 0.2121$	0.38	1
Poisson–binomial ($n = 5$)	$\hat{v} = 0.7956$		$\hat{p} = 0.1624$	0.32	1
Poisson–negative binomial	could not be fitted				

[a] denotes X^2 statistic significant at 5% level.

Table 7.4. Observed and expected distributions of population in Ljubljana, Yugoslavia.

Number of units per cell[a]	Number of cells observed	Expected frequency				
		B.	P.	N.B.	N.T.A.	P.B.(5)
0	90	71.20	75.49	88.37	90.10	90.23
1	28	54.83	48.75	33.41	29.53	28.69
2	15	15.84	15.74	13.24	14.85	15.68
3	9	2.03	3.39	5.33	6.08	6.17
4+	2	0.10	0.63	3.65	3.45	3.23
Total number of cells = 144		X^2 = 21.69	13.60	1.59	0.31	0.32
Total number of units = 93		$P_{0.05}$ = 3.84	3.84	3.84	3.84	3.84

[a] Each unit represents 500 people.

useful analogy of this process by citing methods used in the making of sandpaper "in which an electric charge is imparted to the abrasive particles and an opposite charge to the adhesive-coated paper. The particles are individually attracted to the paper but repelled by each other. The result is that they distribute themselves over the paper in an exceedingly uniform pattern". Hence the influence of the spatial distribution of purchasing power must be treated separately, but not independently, from the repelling influence of rival establishments. This aspect is taken up in chapter 8, which introduces bivariate quadrat models that treat both variables simultaneously.

7.3.2 Interurban comparisons

In order to compare the spatial dispersion of retail establishments in Ljubljana with comparable data for another study area, let us turn to data collected in 1967 by the State Department of Employment in California for San Francisco within the 3000 m by 3000 m study area outlined in figure 7.3. The two sets of data are not exactly comparable for two reasons: (1) the classification system used by the Department of Employment (the Standard Industrial Classification) was not identical to that used in Yugoslavia, and (2) the location of establishments in San Francisco was referenced by block numbers and not by coordinates. Hence, of necessity, we have assigned to all establishments in a single block the coordinates of that block's 'center of gravity'. In figure 7.4, however, the establishments are located side by side, and not on top of one another, for purposes of illustration. The study area, once again is gridded into 144 square quadrats, each measuring 250 m on a side. Also, as in Ljubljana, the study area is centered on the city's central business district. Table 7.5 presents the fits of seven clustered distributions to the

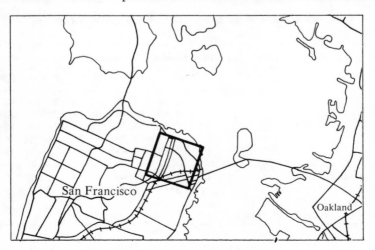

Figure 7.4. Location of the sample grid in San Francisco.

data on the spatial dispersion of San Francisco's 251 food stores and 221 apparel stores.

As in Ljubljana, the spatial dispersion of apparel stores in San Francisco can be satisfactorily described by the negative binomial distribution. Moreover, it appears that the degree of clustering of these stores is roughly the same in the two study areas. For, although the mean number of stores per quadrat in San Francisco is about three times as large as the corresponding figure for Ljubljana, the estimates of the exponent k are almost identical:

Ljubljana
(68 stores)
$\hat{w} = 0.4722$
$\hat{k} = 0.1038$

San Francisco
(221 stores)
$\hat{w} = 1.5347$
$\hat{k} = 0.1024$.

Table 7.5. Parameter estimates and the goodness of fit for food and apparel establishments in San Francisco, USA.

a. Food stores

Model	$\hat{m}_1 = 1.7431$	$\hat{m}_2 = 8.4860$	$\hat{m}_2/\hat{m}_1 = 4.8684$	X^2	d.f.
Negative binomial	$\hat{w} = 1.7431$		$\hat{k} = 0.5988$	14.09[a]	4
Neyman Type A	$\hat{v} = 1.1822$		$\hat{a} = 1.4745$	28.25[a]	4
Poisson–binomial ($n=2$)	$\hat{v} = 1.3498$		$\hat{p} = 0.6457$	34.98[a]	3
Poisson–binomial ($n=3$)	$\hat{v} = 1.2522$		$\hat{p} = 0.4640$	34.80[a]	4
Poisson–binomial ($n=4$)	$\hat{v} = 1.2112$		$\hat{p} = 0.3598$	32.67[a]	4
Poisson–binomial ($n=5$)	$\hat{v} = 1.1949$		$\hat{p} = 0.2917$	31.65[a]	4
Poisson–negative binomial	$\hat{v}_1 = 8.8133$	$\hat{k}_1 = 0.0539$	$\hat{p}_1 = 3.6707$	21.03[a]	3

b. Apparel stores

Model	$\hat{m}_1 = 1.5347$	$\hat{m}_2 = 27.4394$	$\hat{m}_2/\hat{m}_1 = 17.8790$	X^2	d.f.
Negative binomial	$\hat{w} = 1.5347$		$\hat{k} = 0.1024$	3.99	3
Neyman Type A	$\hat{v} = 0.3896$		$\hat{a} = 3.9397$	32.48[a]	4
Poisson–binomial ($n=2$)	$\hat{v} = 0.8482$		$\hat{p} = 0.9047$	74.78[a]	3
Poisson–binomial ($n=3$)	$\hat{v} = 0.6443$		$\hat{p} = 0.7939$	38.48[a]	2
Poisson–binomial ($n=4$)	$\hat{v} = 0.5492$		$\hat{p} = 0.6986$	36.93[a]	3
Poisson–binomial ($n=5$)	$\hat{v} = 0.4993$		$\hat{p} = 0.6147$	31.05[a]	3
Poisson–negative binomial	$\hat{v}_1 = 0.1766$	$\hat{k}_1 = 1.0613$	$\hat{p}_1 = 8.1885$	59.68[a]	1

$\hat{v}_1, \hat{k}_1, \hat{p}_1$ are moment estimates of $v, k,$ and p.
[a] denotes X^2 statistic significant at 5% level.

Observe, however, that unlike the data for Ljubljana, the spatial dispersion of food stores in San Francisco is not satisfactorily accounted for by any of the quadrat distributions that were used for the Ljubljana data. An examination of the various frequency distributions indicates that we need one which has more cells containing one store, while at the same time having a large number of empty quadrats. This suggests that a combination of the negative binomial and the positive binomial, such as the negative binomial–binomial generalized distribution (negative binomial $\vee$ binomial), might provide a satisfactory fit.

It is of interest to compare these findings with comparable ones found for Artle's (1965) data on five categories of retail establishments in downtown Stockholm. Using the same quadrat size, Artle calculated an 'index of association' by dividing the standard deviation of each empirical frequency distribution by that of the corresponding Poisson distribution and then multiplying the quotient by 100. Thus an index of 100 would indicate random dispersion, one lower than that would suggest regular dispersion, and an index larger than 100 would indicate clustered dispersion. Among the five clustered classes of retail establishments, the ordering from highest to lowest clustering was as follows:

	Index of association
Antique stores	248
Women's clothing (apparel) stores	227
Furniture stores and floor covering stores	187
Grocery stores	143
Tobacconists	132 .

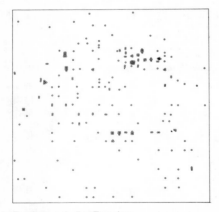

Food stores in San Francisco   Apparel stores in San Francisco

Figure 7.5. The spatial patterns of food and apparel stores in San Francisco, USA.

Since Artle included data on the observed frequency distributions of these five classes of retail establishments, we may fit our quadrat models to them to see if our results confirm those of Artle. Table 7.6 presents the appropriate parameter estimates and goodness of fit statistics. The negative binomial distribution fits four of the five categories of retail trade examined and 'almost' fits the fifth—antique stores. Ranking these from the highest to the lowest clustering on the basis of the estimated parameter k in the negative binomial distribution, we obtain the following values:

	$\hat{k}$
Antique stores	0·1047
Women's clothing (apparel) stores	0·3050
Furniture stores and floor covering stores	0·4077
Grocery stores	1·2662
Tobacconists	3·3644 .

Thus our findings parallel those of Artle.

Table 7.6. Parameter estimates and the goodness of fit for various classes of retail establishments in Stockholm, Sweden.

a. Antique stores

Model	$\hat{m}_1 = 0·4762$	$\hat{m}_2 = 3·2076$	$\hat{m}_2/\hat{m}_1 = 6·7359$	X^2	d.f.
Negative binomial	$\hat{w} = 0·4762$		$\hat{k} = 0·1047$	6·83*	2
Neyman Type A	$\hat{v} = 0·2367$		$\hat{a} = 2·0120$	16·42*	2
Poisson-binomial ($n = 2$)	$\hat{v} = 0·3341$		$\hat{p} = 0·7127$	20·60*	1
Poisson-binomial ($n = 3$)	$\hat{v} = 0·2802$		$\hat{p} = 0·5666$	14·01*	1
Poisson-binomial ($n = 4$)	$\hat{v} = 0·2727$		$\hat{p} = 0·4366$	19·98*	2
Poisson-binomial ($n = 5$)	$\hat{v} = 0·2597$		$\hat{p} = 0·3668$	18·63*	2
Poisson-negative binomial	$\hat{v}_1 = 0·2771$	$\hat{k}_1 = 0·4277$	$\hat{p}_1 = 4·0176$	19·38*	2

b. Women's clothing (apparel) stores

Model	$\hat{m}_1 = 1·0810$	$\hat{m}_2 = 5·5867$	$\hat{m}_2/\hat{m}_1 = 5·1683$	X^2	d.f.
Negative binomial	$\hat{w} = 1·0810$		$\hat{k} = 0·3050$	5·80	4
Neyman Type A	$\hat{v} = 0·5503$		$\hat{a} = 1·9641$	5·14	4
Poisson-binomial ($n = 2$)	$\hat{v} = 0·7153$		$\hat{p} = 0·7556$	27·57*	2
Poisson-binomial ($n = 3$)	$\hat{v} = 0·6054$		$\hat{p} = 0·5952$	15·19*	3
Poisson-binomial ($n = 4$)	$\hat{v} = 0·5653$		$\hat{p} = 0·4780$	10·65*	3
Poisson-binomial ($n = 5$)	$\hat{v} = 0·5514$		$\hat{p} = 0·3921$	10·17*	4
Poisson-negative binomial	$\hat{v} = 7·2595$	$\hat{k} = 0·0447$	$\hat{p} = 3·3303$	5·49	3

Table 7.6 (continued)

c. Furniture stores and floor covering stores

Model	$\hat{m}_1 = 1 \cdot 0095$	$\hat{m}_2 = 3 \cdot 5214$	$\hat{m}_2/\hat{m}_1 = 3 \cdot 4882$	X^2	d.f.
Negative binomial	$\hat{w} = 1 \cdot 0095$		$\hat{k} = 0 \cdot 4077$	5·50	4
Neyman Type A	$\hat{v} = 0 \cdot 6882$		$\hat{a} = 1 \cdot 4668$	11·06*	3
Poisson-binomial ($n = 2$)	$\hat{v} = 0 \cdot 8037$		$\hat{p} = 0 \cdot 6280$	23·38*	2
Poisson-binomial ($n = 3$)	$\hat{v} = 0 \cdot 6769$		$\hat{p} = 0 \cdot 4972$	19·99*	3
Poisson-binomial ($n = 4$)	$\hat{v} = 0 \cdot 6622$		$\hat{p} = 0 \cdot 3811$	17·03*	3
Poisson-binomial ($n = 5$)	$\hat{v} = 0 \cdot 6622$		$\hat{p} = 0 \cdot 3049$	15·43*	3
Poisson-negative binomial	$\hat{v}_1 = 1 \cdot 4048$	$\hat{k}_1 = 0 \cdot 4061$	$\hat{p}_1 = 1 \cdot 7696$	7·16	3

d. Grocery stores

Model	$\hat{m}_1 = 1 \cdot 7143$	$\hat{m}_2 = 3 \cdot 6787$	$\hat{m}_2/\hat{m}_1 = 2 \cdot 1459$	X^2	d.f.
Negative binomial	$\hat{w} = 1 \cdot 7143$		$\hat{k} = 1 \cdot 2662$	4·62	4
Neyman Type A	$\hat{v} = 1 \cdot 5266$		$\hat{a} = 1 \cdot 1229$	1·86	5
Poisson-binomial ($n = 2$)	$\hat{v} = 1 \cdot 2851$		$\hat{p} = 0 \cdot 6670$	8·60	4
Poisson-binomial ($n = 3$)	$\hat{v} = 1 \cdot 2754$		$\hat{p} = 0 \cdot 4480$	3·52	4
Poisson-binomial ($n = 4$)	$\hat{v} = 1 \cdot 3201$		$\hat{p} = 0 \cdot 3247$	2·66	4
Poisson-binomial ($n = 5$)	$\hat{v} = 1 \cdot 3534$		$\hat{p} = 0 \cdot 2533$	2·33	4
Poisson-negative binomial	$\hat{v}_2 = 1 \cdot 5626$	$\hat{k}_2 = 22 \cdot 4500$	$\hat{p}_2 = 0 \cdot 0489$	1·85	4

e. Tobacconists

Model	$\hat{m}_1 = 2 \cdot 9571$	$\hat{m}_2 = 5 \cdot 1226$	$\hat{m}_2/\hat{m}_1 = 1 \cdot 7323$	X^2	d.f.
Negative binomial	$\hat{w} = 2 \cdot 9571$		$\hat{k} = 3 \cdot 3644$	6·97	6
Neyman Type A	$\hat{v} = 3 \cdot 3327$		$\hat{a} = 0 \cdot 8873$	3·62	6
Poisson-binomial ($n = 2$)	$\hat{v} = 2 \cdot 2560$		$\hat{p} = 0 \cdot 6554$	4·82	6
Poisson-binomial ($n = 3$)	$\hat{v} = 2 \cdot 3933$		$\hat{p} = 0 \cdot 4119$	2·81	6
Poisson-binomial ($n = 4$)	$\hat{v} = 2 \cdot 5930$		$\hat{p} = 0 \cdot 2851$	2·91	6
Poisson-binomial ($n = 5$)	$\hat{v} = 2 \cdot 7285$		$\hat{p} = 0 \cdot 2168$	3·03	6
Poisson-negative binomial	could not be fitted				

$\hat{v}_1, \hat{k}_1, \hat{p}_1$ are moment estimates of v, k, and p.
$\hat{v}_2, \hat{k}_2, \hat{p}_2$ are moment-ratio of first two frequencies estimates of v, k, and p.
* denotes X^2 significant at 5% level.

The parameter k in the negative binomial distribution for apparel stores in Ljubljana and in San Francisco was about $0\cdot10$, yet in Stockholm it turned out to be roughly three times that number. This difference appears to be a consequence of Artle's sampling scheme. Although he used the same quadrat size, Artle considered only those quadrats that met the following two restrictions:
1 Each quadrat should have either a resident or a working population of at least 500 people;
2 The total resident and working population of each quadrat should be not less than 1000 people.

Of the 399 quadrats inside of the 21 by 19 grid system that circumscribed his study area, only 210 satisfied the above two restrictions. Were we to include in our analysis the 189 that did not, the observed frequency distribution would become much more highly skewed, and the estimate of k would decrease considerably. For example, assume that the additional 189 quadrats were empty. Then our observed frequency distribution would have 322 empty cells instead of 123, and our maximum likelihood estimate of k would be $0\cdot1274$ instead of $0\cdot3050$.

7.4 The problem of optimal quadrat size

In quadrat analysis the most commonly used two-dimensional areal unit of fixed size and shape is the square. The square has great practical advantages in that it may be used to cover a study region exhaustively in the form of a grid. And it also offers the computational advantage of simplicity. From time to time, however, ecologists have pointed out that the size of the square which is used has a considerable impact on the findings that are generated by quadrat analysis (Greig-Smith, 1964). They argue that the appearance of nonrandomness in a sample of quadrats is not an absolute characteristic, but depends on the size and shape of the sample unit used. Thus, for example, a clustered pattern sampled with a relatively small quadrat will generally yield findings that are quite different from those that would be indicated by a sample drawn using a relatively large quadrat. A very small quadrat will tend to be unoccupied more frequently than not, and no quadrat will contain a large number of points. Hence nonrandomness may not be detected unless the number of points and the number of quadrats is large. Similarly, if a very large quadrat is used such that it includes entire clusters, the distribution of points will tend to differ only within, and not between, quadrats. Thus once again nonrandomness may not be identified. For some intermediate size of quadrat, however, we shall find a relatively large proportion of unoccupied cells and also several quadrats containing many points. In such instances there will be clear evidence of nonrandomness.

The shortcomings of the quadrat method have become increasingly apparent in recent years as the ways in which quadrat size affects spatial data have been explored much **more precisely.** Much of the ecological

Table 7.7. Relation of frequency distribution to quadrat size for five classes of retail establishments in Ljubljana, Yugoslavia.

Number of stores per quadrat	Quadrat size														
	$(50 \text{ m})^2$					$(100 \text{ m})^2$					$(150 \text{ m})^2$				
	Food and nonfood stores	Food stores	Nonfood stores	Grocery stores	Apparel stores	Food and nonfood stores	Food stores	Nonfood stores	Grocery stores	Apparel stores	Food and nonfood stores	Food stores	Nonfood stores	Grocery stores	Apparel stores
0	3341	3435	3457	3522	3547	744	776	811	828	862	286	304	335	341	368
1	154	141	91	77	43	67	78	42	69	25	42	50	28	43	21
2	65	17	28	1	6	33	30	16	5	6	22	20	14	12	5
3	20	4	15		3	17	8	7		3	15	11	2	4	1
4	12	1	4		1	10	4	7		2	9	10	3		0
5	4	2	2			7	2	8		1	7	0	4		2
6	3		2			7	1	4			3	3	3		0
7	0		0			6	1	2			0	0	3		2
8	0		0			3		0			3	1	1		0
9	0		1			2		0		1	2	0	3		1
10	1					1		0			1	0	1		1
11						0		1			1	0	0		
12						1		1			0	0	2		
13						0		1			2	1	1		
14						1		0			0		1		
15						0		0			2		0		
16						0		0			0		0		
17						0		1			2		1		
18						0		0			0		0		
19						1		0			0		0		
20+											3		1		
Total no. of quadrats	3600	3600	3600	3600	3600	900	900	900	900	900	400	400	400	400	400

Table 7.7 (continued)

Number of stores per quadrat	(200 m)²					(250 m)²					(300 m)²				
	Food and nonfood stores	Food stores	Nonfood stores	Grocery stores	Apparel stores	Food and nonfood stores	Food stores	Nonfood stores	Grocery stores	Apparel stores	Food and nonfood stores	Food stores	Nonfood stores	Grocery stores	Apparel stores
0	141	150	179	173	199	78	83	106	96	120	45	49	64	59	81
1	24	26	16	34	14	16	18	15	33	15	12	13	13	25	9
2	16	21	8	12	5	12	13	6	8	3	7	11	5	7	3
3	9	12	3	4	2	9	9	2	1	2	7	6	7	4	3
4	8	7	2	1	1	4	7	2	3	1	5	7	1	1	2
5	4	2	2	1		5	7	1	3		6	5	0	3	
6	6	4	3		1	3	2	2		1	0	2	2	0	0
7	2	1	0			2	1	0			4	0	1	0	
8	3	0	0		1	2	2	1			2	0	0	1	
9	1	1	3			1	0	0			1	5	0		
10	0	0	0			0	1	0			1	0	1		
11	1	0	2			2	0	2			2	0	0		
12	0	0	0			1	0	0			1	0	0		
13	1	0	2			0	0	2			0	0	0		
14	1	0	0			0	0	1			1	0	0		
15	1	0	1		1	0	0	0		1	0	0	0		
16	0	0	0			2	0	1			0	0	1		
17	0	0	0			0	0	0			0	0	0		
18	0	0	0			2	0	0			0	2	0		
19	2	1	0			0	1	0			0	0	0		
20+	4	2	2			5	1	2			5	2	3		1
Total no. of quadrats	225	225	225	225	225	144	144	144	144	144	100	100	100	100	100

literature on this subject has been reviewed by Curtis and McIntosh (1950), who suggest the general rule of thumb that the appropriate size of square is approximately twice the size of the mean area per point. Greig-Smith (1964), on the other hand, suggests that minimizing the sample variance and striving for a reasonably symmetric distribution are appropriate criteria for certain classes of distributions. In this section we outline our approach for determining the optimal quadrat size—one which we feel is more in keeping with the Neyman–Pearson theory of statistical inference. In particular we propose that *the optimal quadrat size is that size which maximizes the power of the chi-square test for a given level of significance*, α. This criterion, of course, requires that one should specify not only the null hypothesis, but also the alternative hypothesis as well.

As an illustration consider the data on the spatial dispersion of retail establishments in Ljubljana, Yugoslavia that appear in table 7.7. The total number of food and nonfood establishments is 440, and for six different quadrat sizes—$(50\text{ m})^2$, $(100\text{ m})^2$, $(150\text{ m})^2$, $(200\text{ m})^2$, $(250\text{ m})^2$, and $(300\text{ m})^2$—we have the following sample estimates of $\hat{m}_1$ and $\hat{m}_2/\hat{m}_1$:

$\hat{m}_1$ 0·12   0·49   1·10   1·96   3·06   4·40
$\hat{m}_2/\hat{m}_1$ 2·36   4·94   8·38   12·29   17·36   25·01 .

By plotting these points in figure 5.2, we see that the power is 1·0 for all but the smallest quadrat size.

The total number of food stores in Ljubljana is 201, and we have, for the six different sizes of cells, the following values:

$\hat{m}_1$ 0·06   0·22   0·50   0·89   1·40   2·01
$\hat{m}_2/\hat{m}_1$ 1·49   2·11   3·01   4·18   4·96   5·74 .

If we plot these data in figure 5.3, we find that the power increases from a minimum of 0·33 for $\hat{m}_1 = 0·06$ to a maximum of 1·00 for $\hat{m}_1 = 0·89$, and then declines to 0·70 for $\hat{m}_1 = 2·01$.

As a final example we consider the spatial pattern of grocery stores. There are 79 such stores in the study area and, for the six different grid sizes, we obtain:

$\hat{m}_1$ 0·02   0·09   0·20   0·35   0·55   0·79
$\hat{m}_2/\hat{m}_1$ 1·00   1·04   1·41   1·67   1·96   2·54 .

Using figure 5.4, we determine that the power increases monotonically with quadrat size and that the largest quadrat size yields the maximum power of 0·14.

Finally, in table 7.8, we present the fits of both the Neyman Type A and the negative binomial distributions to the Ljubljana data. Note that for total stores (food and nonfood) discriminating between the two distributions is not difficult since only the negative binomial provides a satisfactory fit. For food stores, however, the discrimination problem begins to emerge as both distributions are not rejected in four of the six cases. (Observe, however, that the 200 m by 200 m quadrat size does provide the best fitting Neyman Type A distribution.) The problem of discrimination reaches its peak in the case of grocery stores. Here it is virtually impossible to choose between the two distributions. Recall that this decline in discriminatory ability is precisely what was indicated by the power functions. It is interesting to note, however, that despite a smaller number of stores—that is, 68—the power for apparel stores is quite high and the problem of discrimination does not appear. The same holds true for nonfood stores. For these two classes of retail establishments, the Neyman Type A distribution is rejected in all but a single case.

Table 7.8. Effect of quadrat size on the variance-mean ratio, the parameter estimates, and the goodness of fit for food and nonfood, food, grocery, nonfood, and apparel stores in Ljubljana, Yugoslavia.

	Model	Quadrat size (m)^2			X^2	d.f.
Food and nonfood			$\hat{m}_1 = 0\cdot1222$	$\hat{m}_2/\hat{m}_1 = 2\cdot3648$		
	N.B.	50	$\hat{w} = 0\cdot1222$	$\hat{k} = 0\cdot0820$	4·33	4
	N.T.A.		$\hat{v} = 0\cdot1156$	$\hat{a} = 1\cdot0571$	11·21[a]	2
			$\hat{m}_1 = 0\cdot4889$	$\hat{m}_2/\hat{m}_1 = 4\cdot9439$		
	N.B.	100	$\hat{w} = 0\cdot4889$	$\hat{k} = 0\cdot1150$	2·59	6
	N.T.A.		$\hat{v} = 0\cdot2326$	$\hat{a} = 2\cdot1017$	50·44[a]	4
			$\hat{m}_1 = 1\cdot1000$	$\hat{m}_2/\hat{m}_1 = 8\cdot3800$		
	N.B.	150	$\hat{w} = 1\cdot1000$	$\hat{k} = 0\cdot1626$	4·20	6
	N.T.A.		$\hat{v} = 0\cdot4105$	$\hat{a} = 2\cdot6796$	43·00[a]	5
			$\hat{m}_1 = 1\cdot9556$	$\hat{m}_2/\hat{m}_1 = 12\cdot2945$		
	N.B.	200	$\hat{w} = 1\cdot9556$	$\hat{k} = 0\cdot1927$	1·99	7
	N.T.A.		$\hat{v} = 0\cdot5750$	$\hat{a} = 3\cdot4007$	51·80[a]	7
			$\hat{m}_1 = 3\cdot0556$	$\hat{m}_2/\hat{m}_1 = 17\cdot3605$		
	N.B.	250	$\hat{w} = 3\cdot0556$	$\hat{k} = 0\cdot2255$	4·41	6
	N.T.A.		$\hat{v} = 0\cdot7765$	$\hat{a} = 3\cdot9351$	46·27[a]	7
			$\hat{m}_1 = 4\cdot4000$	$\hat{m}_2/\hat{m}_1 = 25\cdot0092$		
	N.B.	300	$\hat{w} = 4\cdot4000$	$\hat{k} = 0\cdot2743$	5·02	6
	N.T.A.		$\hat{v} = 0\cdot9920$	$\hat{a} = 4\cdot4357$	39·03[a]	7

[a] denotes X^2 statistic significant at 5% level.

Table 7.8 (continued)

	Model	Quadrat size (m)^2			X^2	d.f.
Food			$\hat{m}_1 = 0.0558$	$\hat{m}_2/\hat{m}_1 = 1.4918$		
	N.B.	50	$\hat{w} = 0.0558$	$\hat{k} = 0.1366$	1.65	1
	N.T.A.		$\hat{v} = 0.1614$	$\hat{a} = 0.3459$	(5.34[a])	1
			$\hat{m}_1 = 0.2233$	$\hat{m}_2/\hat{m}_1 = 2.1124$		
	N.B.	100	$\hat{w} = 0.2233$	$\hat{k} = 0.1935$	1.21	2
	N.T.A.		$\hat{v} = 0.2608$	$\hat{a} = 0.8564$	4.32	2
			$\hat{m}_1 = 0.5025$	$\hat{m}_2/\hat{m}_1 = 3.0125$		
	N.B.	150	$\hat{w} = 0.5025$	$\hat{k} = 0.2487$	4.44	3
	N.T.A.		$\hat{v} = 0.3877$	$\hat{a} = 1.2960$	7.30[a]	2
			$\hat{m}_1 = 0.8933$	$\hat{m}_2/\hat{m}_1 = 4.1750$		
	N.B.	200	$\hat{w} = 0.8933$	$\hat{k} = 0.2975$	5.19	4
	N.T.A.		$\hat{v} = 0.4995$	$\hat{a} = 1.7883$	1.56	3
			$\hat{m}_1 = 1.3958$	$\hat{m}_2/\hat{m}_1 = 4.9620$		
	N.B.	250	$\hat{w} = 1.3958$	$\hat{k} = 0.3544$	3.99	4
	N.T.A.		$\hat{v} = 0.6435$	$\hat{a} = 2.1691$	4.14	4
			$\hat{m}_1 = 2.0100$	$\hat{m}_2/\hat{m}_1 = 5.7440$		
	N.B.	300	$\hat{w} = 2.0100$	$\hat{k} = 0.4130$	6.08	4
	N.T.A.		$\hat{v} = 0.8214$	$\hat{a} = 2.4469$	4.82	4
Grocery			$\hat{m}_1 = 0.0219$	$\hat{m}_2/\hat{m}_1 = 1.0037$		
	N.B.	50	$\hat{w} = 0.0219$	$\hat{k} = 6.3980$	–	0
	N.T.A.		$\hat{v} = 6.3101$	$\hat{a} = 0.0035$	–	0
			$\hat{m}_1 = 0.0878$	$\hat{m}_2/\hat{m}_1 = 1.0400$		
	N.B.	100	$\hat{w} = 0.0878$	$\hat{k} = 2.0718$	–	0
	N.T.A.		$\hat{v} = 1.9720$	$\hat{a} = 0.0445$	–	0
			$\hat{m}_1 = 0.1975$	$\hat{m}_2/\hat{m}_1 = 1.4136$		
	N.B.	150	$\hat{w} = 0.1975$	$\hat{k} = 0.4017$	(0.35)	1
	N.T.A.		$\hat{v} = 0.4434$	$\hat{a} = 0.4454$	(0.10)	1
			$\hat{m}_1 = 0.3511$	$\hat{m}_2/\hat{m}_1 = 1.6690$		
	N.B.	200	$\hat{w} = 0.3511$	$\hat{k} = 0.4853$	0.14	1
	N.T.A.		$\hat{v} = 0.5821$	$\hat{a} = 0.6031$	0.14	1
			$\hat{m}_1 = 0.5486$	$\hat{m}_2/\hat{m}_1 = 1.9587$		
	N.B.	250	$\hat{w} = 0.5486$	$\hat{k} = 0.6374$	1.49	1
	N.T.A.		$\hat{v} = 0.8975$	$\hat{a} = 0.6112$	2.61	1
			$\hat{m}_1 = 0.7900$	$\hat{m}_2/\hat{m}_1 = 2.5392$		
	N.B.	300	$\hat{w} = 0.7900$	$\hat{k} = 0.6242$	1.50	1
	N.T.A.		$\hat{v} = 0.9409$	$\hat{a} = 0.8396$	3.25	1

[a] denotes X^2 statistic significant at 5% level.
(X^2) is the X^2 statistic computed with grouping ≥ 1 instead of ≥ 5.

Table 7.8 (continued)

	Model	Quadrat size (m)²			X^2	d.f.
Nonfood			$\hat{m}_1 = 0{\cdot}0664$	$\hat{m}_2/\hat{m}_1 = 2{\cdot}4657$		
	N.B.	50	$\hat{w} = 0{\cdot}0664$	$\hat{k} = 0{\cdot}0442$	1·30	3
	N.T.A.		$\hat{v} = 0{\cdot}0628$	$\hat{a} = 1{\cdot}0564$	8·44[a]	2
			$\hat{m}_1 = 0{\cdot}2655$	$\hat{m}_2/\hat{m}_1 = 4{\cdot}8750$		
	N.B.	100	$\hat{w} = 0{\cdot}2655$	$\hat{k} = 0{\cdot}0632$	5·91	4
	N.T.A.		$\hat{v} = 0{\cdot}1224$	$\hat{a} = 2{\cdot}1345$	32·33[a]	3
			$\hat{m}_1 = 0{\cdot}5975$	$\hat{m}_2/\hat{m}_1 = 8{\cdot}2893$		
	N.B.	150	$\hat{w} = 0{\cdot}5975$	$\hat{k} = 0{\cdot}0844$	4·95	4
	N.T.A.		$\hat{v} = 0{\cdot}2179$	$\hat{a} = 2{\cdot}7420$	64·53[a]	4
			$\hat{m}_1 = 1{\cdot}0622$	$\hat{m}_2/\hat{m}_1 = 11{\cdot}2934$		
	N.B.	200	$\hat{w} = 1{\cdot}0622$	$\hat{k} = 0{\cdot}0894$	1·52	3
	N.T.A.		$\hat{v} = 0{\cdot}2903$	$\hat{a} = 3{\cdot}6591$	29·50[a]	4
			$\hat{m}_1 = 1{\cdot}6597$	$\hat{m}_2/\hat{m}_1 = 17{\cdot}8997$		
	N.B.	250	$\hat{w} = 1{\cdot}6597$	$\hat{k} = 0{\cdot}1073$	3·49	3
	N.T.A.		$\hat{v} = 0{\cdot}4205$	$\hat{a} = 3{\cdot}9475$	39·84[a]	5
			$\hat{m}_1 = 2{\cdot}3900$	$\hat{m}_2/\hat{m}_1 = 24{\cdot}4275$		
	N.B.	300	$\hat{w} = 2{\cdot}3900$	$\hat{k} = 0{\cdot}1524$	3·62	3
	N.T.A.		$\hat{v} = 0{\cdot}6075$	$\hat{a} = 3{\cdot}9341$	30·13[a]	4
Apparel			$\hat{m}_1 = 0{\cdot}0189$	$\hat{m}_2/\hat{m}_1 = 1{\cdot}5992$		
	N.B.	50	$\hat{w} = 0{\cdot}0189$	$\hat{k} = 0{\cdot}0318$	(1·01)	1
	N.T.A.		$\hat{v} = 0{\cdot}0377$	$\hat{a} = 0{\cdot}5017$	(3·73)	1
			$\hat{m}_1 = 0{\cdot}0756$	$\hat{m}_2/\hat{m}_1 = 3{\cdot}0749$		
	N.B.	100	$\hat{w} = 0{\cdot}0756$	$\hat{k} = 0{\cdot}0416$	0·55	1
	N.T.A.		$\hat{v} = 0{\cdot}0636$	$\hat{a} = 1{\cdot}1873$	4·35[a]	1
			$\hat{m}_1 = 0{\cdot}1700$	$\hat{m}_2/\hat{m}_1 = 4{\cdot}1639$		
	N.B.	150	$\hat{w} = 0{\cdot}1700$	$\hat{k} = 0{\cdot}0637$	1·85	1
	N.T.A.		$\hat{v} = 0{\cdot}1158$	$\hat{a} = 1{\cdot}4679$	7·22[a]	1
			$\hat{m}_1 = 0{\cdot}3022$	$\hat{m}_2/\hat{m}_1 = 5{\cdot}8709$		
	N.B.	200	$\hat{w} = 0{\cdot}3022$	$\hat{k} = 0{\cdot}0760$	0·58	1
	N.T.A.		$\hat{v} = 0{\cdot}1566$	$\hat{a} = 1{\cdot}9300$	7·26[a]	2
			$\hat{m}_1 = 0{\cdot}4722$	$\hat{m}_2/\hat{m}_1 = 8{\cdot}0839$		
	N.B.	250	$\hat{w} = 0{\cdot}4722$	$\hat{k} = 0{\cdot}1038$	3·53	1
	N.T.A.		$\hat{v} = 0{\cdot}2426$	$\hat{a} = 1{\cdot}9464$	12·31[a]	2
			$\hat{m}_1 = 0{\cdot}6800$	$\hat{m}_2/\hat{m}_1 = 12{\cdot}5336$		
	N.B.	300	$\hat{w} = 0{\cdot}6800$	$\hat{k} = 0{\cdot}1011$	1·18	1
	N.T.A.		$\hat{v} = 0{\cdot}2659$	$\hat{a} = 2{\cdot}5570$	(11·64[a])	5

[a] denotes X^2 statistic significant at 5% level.
(X^2) is the X^2 statistic computed with grouping $\geqslant 1$ instead of $\geqslant 5$.

7.5 The spatial dispersion of retail employment, space, and sales in Ljubljana, Yugoslavia

In chapter 6 we concluded that the size distribution of retail trade in Ljubljana is highly concentrated, with a relatively few establishments accounting for a large proportion of the study area's total retail employment, space, and sales. In this chapter we have concluded that the spatial dispersion of the same retail establishments is highly clustered, with a relatively few quadrats accounting for a large proportion of the study area's total population of retail shops. We would expect therefore to find that the spatial dispersion of the study area's retail employment, space, and sales is even more highly clustered than the spatial dispersion of its retail establishments. And quadrat analysis of our data reveals that this indeed is the case.

Table 7.9 presents the results of a quadrat analysis of the spatial dispersion of food store employment in Ljubljana. Table 7.9a reports the results obtained with a quadrat size of 250 m by 250 m and an interval width of 5 employees. To examine the effects of a wider interval, we repeated the analysis with an interval of 10 employees; and to test the impact of a change in quadrat size, we also examined the effects of reducing our quadrat size to 150 m by 150 m. These two sets of results are summarized in tables 7.9b and 7.9c respectively.

It appears that the negative binomial distribution provides a satisfactory fit to the data. However, note the sensitivity of the results to changes in interval width and quadrat size. Increasing the width of the interval increases clustering, as does a decrease in the quadrat size.

In table 7.10 we present parallel results for nonfood stores. Holding the quadrat size constant at 150 m by 150 m, but now including retail space and retail sales as additional measures of the level of retail trade, we find that none of our quadrat models fits the data[19]. The principal reason for this is the insufficient degree of skewness exhibited by our theoretical distributions. Nonfood stores are spatially very clustered, and the introduction of their skewed size distributions into the spatial analysis results in spatial dispersions that can only be accounted for by compound and generalized distributions lying to the right of the negative binomial distribution on the dispersion line—for example, the negative binomial $\vee$ negative binomial distribution.

Despite the relatively poor fits of the negative binomial distribution to the data on nonfood store employment, space, and sales, the estimates of the parameter k can still offer us some insights into the relative degrees of spatial clustering exhibited by these three measures of retail trade.

---

[19] The unsatisfactory fits of the negative binomial distribution to the data on nonfood stores persisted across all quadrat sizes up to the 10 by 10 grid system, with cells 300 m on a side, and across several different interval widths.

Table 7.9. The spatial dispersion of food store employment in Ljubljana, Yugoslavia; parameter estimates and the goodness of fit.

a. Interval width = 5 employees; quadrat size = $(250 \text{ m})^2$

Model	$\hat{m}_1 = 1\cdot 6250$	$\hat{m}_2 = 19\cdot 7745$	$\hat{m}_2/\hat{m}_1 = 12\cdot 1689$	X^2	d.f.
Negative binomial	$\hat{w} = 1\cdot 6250$		$\hat{k} = 0\cdot 1826$	$1\cdot 56$	4
Neyman Type A	$\hat{v} = 0\cdot 5271$		$\hat{a} = 3\cdot 0830$	$27\cdot 03^a$	5
Poisson-binomial ($n = 2$)	$\hat{v} = 0\cdot 9632$		$\hat{p} = 0\cdot 8435$	$57\cdot 44^a$	3
Poisson-binomial ($n = 3$)	$\hat{v} = 0\cdot 7607$		$\hat{p} = 0\cdot 7120$	$41\cdot 32^a$	4
Poisson-binomial ($n = 4$)	$\hat{v} = 0\cdot 6552$		$\hat{p} = 0\cdot 6201$	$34\cdot 90^a$	4
Poisson-binomial ($n = 5$)	$\hat{v} = 0\cdot 6187$		$\hat{p} = 0\cdot 5253$	$35\cdot 05^a$	4
Poisson-negative binomial	$\hat{v}_1 = 0\cdot 4896$	$\hat{k}_1 = 0\cdot 4228$	$\hat{p}_1 = 7\cdot 8500$	$22\cdot 59^a$	3

b. Interval width = 10 employees; quadrat size = $(250 \text{ m})^2$

Model	$\hat{m}_1 = 0\cdot 7847$	$\hat{m}_2 = 7\cdot 7785$	$\hat{m}_2/\hat{m}_1 = 9\cdot 9124$	X^2	d.f.
Negative binomial	$\hat{w} = 0\cdot 7847$		$\hat{k} = 0\cdot 1311$	$2\cdot 21$	2
Neyman Type A	$\hat{v} = 0\cdot 3333$		$\hat{a} = 2\cdot 3544$	$12\cdot 66^a$	3
Poisson-binomial ($n = 2$)	$\hat{v} = 0\cdot 4976$		$\hat{p} = 0\cdot 7884$	$20\cdot 34^a$	1
Poisson-binomial ($n = 3$)	$\hat{v} = 0\cdot 4065$		$\hat{p} = 0\cdot 6435$	$15\cdot 92^a$	2
Poisson-binomial ($n = 4$)	$\hat{v} = 0\cdot 3730$		$\hat{p} = 0\cdot 5260$	$14\cdot 62^a$	2
Poisson-binomial ($n = 5$)	$\hat{v} = 0\cdot 3631$		$\hat{p} = 0\cdot 4323$	$13\cdot 54^a$	2
Poisson-negative binomial	$\hat{v}_1 = 1\cdot 1155$	$\hat{k}_1 = 0\cdot 0857$	$\hat{p}_1 = 8\cdot 2090$	$(13\cdot 67)$	7

c. Interval width = 5 employees; quadrat size = $(150 \text{ m})^2$

Model	$\hat{m}_1 = 0\cdot 5725$	$\hat{m}_2 = 5\cdot 4383$	$\hat{m}_2/\hat{m}_1 = 9\cdot 4993$	X^2	d.f.
Negative binomial	$\hat{w} = 0\cdot 5725$		$\hat{k} = 0\cdot 0758$	$4\cdot 96$	4
Neyman Type A	$\hat{v} = 0\cdot 1922$		$\hat{a} = 2\cdot 9794$	$21\cdot 22^a$	4
Poisson-binomial ($n = 2$)	$\hat{v} = 0\cdot 3544$		$\hat{p} = 0\cdot 8076$	$73\cdot 46^a$	2
Poisson-binomial ($n = 3$)	$\hat{v} = 0\cdot 2666$		$\hat{p} = 0\cdot 7158$	$50\cdot 10^a$	2
Poisson-binomial ($n = 4$)	$\hat{v} = 0\cdot 2406$		$\hat{p} = 0\cdot 5948$	$47\cdot 97^a$	3
Poisson-binomial ($n = 5$)	$\hat{v} = 0\cdot 2198$		$\hat{p} = 0\cdot 5210$	$39\cdot 88^a$	3
Poisson-negative binomial	could not be fitted				

$\hat{v}_1, \hat{k}_1, \hat{p}_1$ are moment estimates of v, k, and p.
[a] denotes X^2 statistic significant at 5% level.
(X^2) is the X^2 statistic computed with grouping ≥ 1 instead of ≥ 5.

Table 7.10. The spatial dispersion of nonfood store employment, space, and sales in Ljubljana, Yugoslavia: parameter estimates and the goodness of fit for a quadrat size of 150 m on a side.

a. Employment: interval width = 2 employees

Model	$\hat{m}_1 = 0\cdot9475$	$\hat{m}_2 = 6\cdot9321$	$\hat{m}_2/\hat{m}_1 = 7\cdot3162$	X^2	d.f.
Negative binomial	$\hat{w} = 0\cdot9475$		$\hat{k} = 0\cdot0650$	$17\cdot95^a$	4
Neyman Type A	$\hat{v} = 0\cdot1865$		$\hat{a} = 5\cdot0798$	$93\cdot90^a$	6
Poisson-binomial ($n = 2$)	$\hat{v} = 0\cdot5095$		$\hat{p} = 0\cdot9299$	$322\cdot64^a$	3
Poisson-binomial ($n = 3$)	$\hat{v} = 0\cdot4045$		$\hat{p} = 0\cdot7808$	$193\cdot70^a$	4
Poisson-binomial ($n = 4$)	$\hat{v} = 0\cdot3218$		$\hat{p} = 0\cdot7360$	$173\cdot96^a$	4
Poisson-binomial ($n = 5$)	$\hat{v} = 0\cdot2788$		$\hat{p} = 0\cdot6797$	$111\cdot11^a$	3
Poisson-negative binomial	$\hat{v}_2 = 0\cdot2444$	$\hat{k}_2 = 1\cdot5888$	$\hat{p}_2 = 2\cdot4398$	$27\cdot51^a$	5

b. Space: interval width = $(50\text{ m})^2$

Model	$\hat{m}_1 = 0\cdot7550$	$\hat{m}_2 = 5\cdot7443$	$\hat{m}_2/\hat{m}_1 = 7\cdot6084$	X^2	d.f.
Negative binomial	$\hat{w} = 0\cdot7550$		$\hat{k} = 0\cdot0528$	$18\cdot84^a$	4
Neyman Type A	$\hat{v} = 0\cdot1535$		$\hat{a} = 4\cdot9189$	$71\cdot62^a$	5
Poisson-binomial ($n = 2$)	$\hat{v} = 0\cdot4122$		$\hat{p} = 0\cdot9158$	$120\cdot01^a$	2
Poisson-binomial ($n = 3$)	$\hat{v} = 0\cdot3287$		$\hat{p} = 0\cdot7656$	$101\cdot68^a$	2
Poisson-binomial ($n = 4$)	$\hat{v} = 0\cdot2620$		$\hat{p} = 0\cdot7205$	$127\cdot03^a$	3
Poisson-binomial ($n = 5$)	$\hat{v} = 0\cdot2327$		$\hat{p} = 0\cdot6489$	$119\cdot05^a$	3
Poisson-negative binomial	$\hat{v}_2 = 0\cdot2522$	$\hat{k}_2 = 0\cdot8281$	$\hat{p}_2 = 3\cdot6149$	$35\cdot06^a$	4

c. Sales: interval width = 500 dinars

Model	$\hat{m}_1 = 0\cdot9550$	$\hat{m}_2 = 7\cdot2210$	$\hat{m}_2/\hat{m}_1 = 7\cdot5613$	X^2	d.f.
Negative binomial	$\hat{w} = 0\cdot9550$		$\hat{k} = 0\cdot0571$	$29\cdot59^a$	4
Neyman Type A	$\hat{v} = 0\cdot1609$		$\hat{a} = 5\cdot9350$	$58\cdot92^a$	5
Poisson-binomial ($n = 2$)	$\hat{v} = 0\cdot5084$		$\hat{p} = 0\cdot9393$	$331\cdot82^a$	3
Poisson-binomial ($n = 3$)	$\hat{v} = 0\cdot3966$		$\hat{p} = 0\cdot8026$	$162\cdot06^a$	4
Poisson-binomial ($n = 4$)	$\hat{v} = 0\cdot3166$		$\hat{p} = 0\cdot7541$	$139\cdot34^a$	3
Poisson-binomial ($n = 5$)	$\hat{v} = 0\cdot2296$		$\hat{p} = 0\cdot8317$	$106\cdot40^a$	2
Poisson-negative binomial	$\hat{v}_2 = 0\cdot1961$	$\hat{k}_2 = 2\cdot8772$	$\hat{p}_2 = 1\cdot6923$	$34\cdot02^a$	5

$\hat{v}_2, \hat{k}_2, \hat{p}_2$ are moment-ratio of first two frequencies estimates of $v, k,$ and p.
[a] denotes X^2 significant at 5% level.

And although the results are highly sensitive to decisions made regarding interval width and quadrat size, the orderings seem to persist, with employment always manifesting a less clustered spatial dispersion than either space or sales. Combining this finding with the fact that nonfood retailing is more highly clustered spatially than food retailing suggests the order along the dispersion line shown in figure 7.6.

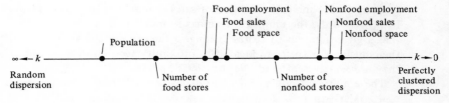

Figure 7.6. Dispersion line for population and retailing in Ljubljana, Yugoslavia.

8

Bivariate distributions

8.1 Introduction
In chapter 7 we used quadrat methods to analyze the spatial dispersion of retail establishments in urban areas. There it was concluded that competing shopping goods establishments tend to cluster together in order to share in the increased volume of sales that such aggregation produces. In other words, the empirical findings supported the marketing principle that a given number of stores dealing in the same merchandise will increase their business if they are located in proximity to each other, rather than if they are widely scattered (Nelson, 1958, p.58).

The empirical findings for convenience goods stores, however, were less clear. Marketing researchers have asserted that such establishments do not profit from cohesion and therefore do not usually group together (Nelson, 1958, p.179). Yet analysis of the data revealed that convenience goods establishments also tended to assume a clustered pattern. This spatial clustering was attributed to the spatial clustering of the residential population. It was argued that, because of the unspecialized and recurrent function that they perform, convenience goods stores tend to be drawn towards the consumers they serve and thus assume a distributional pattern which mirrors that of the residential population. Hence, it was concluded, both spatial patterns must be considered simultaneously.

In this chapter we continue our previous efforts in this direction by introducing bivariate quadrat models which may be used to account for both the number of convenience goods establishments in a cell and the number of people (or purchasing power) present in that same cell. Once again the data used for empirical testing are the spatial patterns of retail establishments and residential population in Ljubljana, Yugoslavia, during the year 1966.

8.2 Random bivariate dispersion: the bivariate correlated Poisson distribution
The notion of 'randomness' has been a fundamental concept throughout this book. Recall its definition and the probabilistic law that was derived to describe it in chapter 1. There we assumed that M points were located randomly in a planar region with λA denoting the probability that a point would fall within a particular subdivision of area A. A square subregion of area a in the plane was subdivided into n very small square subdivisions. It was assumed that these square subdivisions were so small that the probability of more than one point occurring in them was insignificant and tended to zero as n increased.

Let us now introduce subscripts to distinguish two classes of points—type 1 points and type 2 points, say—and assume that no more than one

Bivariate distributions

point of each type may occur in each of the small subdivisions. Then, generalizing the argument developed in subsection 1.2.2, we denote the probability that a subdivision contains *both* a type 1 and a type 2 point by $\lambda_{11} a/n$, the probability that it contains a type 1 but *not* a type 2 point by $\lambda_{10} a/n$, and the probability that it contains a type 2 point but *not* a type 1 point by $\lambda_{01} a/n$. Thus the probability that a subdivision contains no points of either type is $1 - [(\lambda_{11} + \lambda_{10} + \lambda_{01})a/n]$. Therefore it follows that, in a square subregion of area a, the probability of finding x_1 subdivisions with both a single type 1 *and* a single type 2 point, x_2 subdivisions with only a single type 1 point, x_3 subdivisions with only a single type 2 point, and $(n - x_1 - x_2 - x_3)$ subdivisions with no points of either type is [20]

$$P(X_{11} = x_1, X_{10} = x_2, X_{01} = x_3) =$$

$$\binom{n}{x_1 \ x_2 \ x_3} \left(\lambda_{11} \frac{a}{n}\right)^{x_1} \left(\lambda_{10} \frac{a}{n}\right)^{x_2} \left(\lambda_{01} \frac{a}{n}\right)^{x_3}$$

$$\times \left[1 - (\lambda_{11} + \lambda_{10} + \lambda_{01}) \frac{a}{n}\right]^{n - x_1 - x_2 - x_3}.$$

Now, defining R_i to be the random variable denoting the number of type i points in a randomly selected subdivision, we have that $R_1 = X_{11} + X_{10}$, $R_2 = X_{11} + X_{01}$. Whence, for example,

$$P(R_1 = r_1, R_2 = 0) = P(X_{11} = 0) P(X_{10} = r_1) P(X_{01} = 0)$$

$$= \frac{n!}{0! r_1! 0! (n - r_1)!} \left(\lambda_{10} \frac{a}{n}\right)^{r_1} \left[1 - (\lambda_{11} + \lambda_{10} + \lambda_{01}) \frac{a}{n}\right]^{n - r_1},$$

$$P(R_1 = r_1, R_2 = 1) = P(X_{11} = 0) P(X_{10} = r_1) P(X_{01} = 1)$$

$$+ P(X_{11} = 1) P(X_{10} = r_1 - 1) P(X_{01} = 0)$$

$$= \frac{n!}{0! r_1! 1! (n - r_1 - 1)!} \left(\lambda_{10} \frac{a}{n}\right)^{r_1} \left(\lambda_{01} \frac{a}{n}\right)^{1} \left[1 - (\lambda_{11} + \lambda_{10} + \lambda_{01}) \frac{a}{n}\right]^{n - r_1 - 1}$$

$$+ \frac{n!}{1! (r_1 - 1)! 0! (n - r_1)!} \left(\lambda_{11} \frac{a}{n}\right)^{1} \left(\lambda_{10} \frac{a}{n}\right)^{r_1 - 1} \left[1 - (\lambda_{11} + \lambda_{10} + \lambda_{01}) \frac{a}{n}\right]^{n - r_1}$$

$$= \sum_{x=0}^{1} \binom{n}{x \ r_1 - x \ 1 - x} \left(\lambda_{11} \frac{a}{n}\right)^{x} \left(\lambda_{10} \frac{a}{n}\right)^{r_1 - x} \left(\lambda_{01} \frac{a}{n}\right)^{1 - x}$$

$$\times \left[1 - (\lambda_{11} + \lambda_{10} + \lambda_{01}) \frac{a}{n}\right]^{n - r_1 - 1 + x},$$

[20] $\binom{n}{x_1 \ x_2 \ x_3} = \dfrac{n!}{x_1! x_2! x_3! (n - x_1 - x_2 - x_3)!}$.

and
$$P(R_1 = r_1, R_2 = r_2) = \sum_{x=0}^{\min(r_1,r_2)} \binom{n}{x \ \ r_1-x \ \ r_2-x} \qquad (8.1)$$
$$\times \left(\lambda_{11}\frac{a}{n}\right)^x \left(\lambda_{10}\frac{a}{n}\right)^{r_1-x} \left(\lambda_{01}\frac{a}{n}\right)^{r_2-x} \left[1-(\lambda_{11}+\lambda_{10}+\lambda_{01})\frac{a}{n}\right]^{n-r_1-r_2+x},$$

which is the *bivariate correlated binomial* (BCB) distribution with probability generating function

$$G_{12}(s_1,s_2) = \left(1-\lambda_{11}\frac{a}{n}-\lambda_{10}\frac{a}{n}-\lambda_{01}\frac{a}{n}+\lambda_{10}\frac{a}{n}s_1+\lambda_{01}\frac{a}{n}s_2+\lambda_{11}\frac{a}{n}s_1 s_2\right)^n$$

If (8.2)
$\lambda_1 = \lambda_{11}+\lambda_{10}$,
$\lambda_2 = \lambda_{11}+\lambda_{01}$,
then
$\lambda_{10} = \lambda_1 - \lambda_{11}$,
$\lambda_{01} = \lambda_2 - \lambda_{11}$.

Now suppose that, as $n \to \infty$, $\lambda_{11}a/n$, $\lambda_1 a/n$, $\lambda_2 a/n$ all tend to zero in such a manner that

$$n\left(\lambda_1 \frac{a}{n}\right) = v_1, \quad n\left(\lambda_2 \frac{a}{n}\right) = v_2, \quad n\left(\lambda_{11}\frac{a}{n}\right) = v_{12}.$$

We have then (Teicher, 1954) the *bivariate correlated Poisson* (BCP) distribution

$$P(r_1,r_2) = \exp[-(v_1+v_2-v_{12})]\sum_{x=0}^{z} \frac{(v_1-v_{12})^{r_1-x}(v_2-v_{12})^{r_2-x}v_{12}^x}{(r_1-x)!(r_2-x)!x!}, (8.3)$$

where v_1 and v_2 are marginal means, v_{12} is the covariance, and $z = \min(r_1,r_2)$.

A parallel argument carried out on the probability generating function in equation (8.2) gives

$$G_{12}(s_1,s_2) = \exp[v_1(s_1-1)+v_2(s_2-1)+v_{12}(s_1-1)(s_2-1)] \qquad (8.4)$$
$$= \exp[(v_1-v_{12})(s_1-1)+(v_2-v_{12})(s_2-1)+v_{12}(s_1 s_2-1)], (8.5)$$

which is the p.g.f. of the BCP distribution[21].

Teicher (1954) offers the following recurrence relationships which are convenient for calculating the numerical values of $P(r_1,r_2)$:

$$r_1 P(r_1,r_2) = (v_1-v_{12})P(r_1-1,r_2)+v_{12}P(r_1-1,r_2-1),$$
$$r_2 P(r_1,r_2) = (v_2-v_{12})P(r_1,r_2-1)+v_{12}P(r_1-1,r_2-1), \qquad (8.6)$$

for $r_1 \geq 0$, $r_2 \geq 0$.

[21] The BCP distribution also can be derived by assuming that X_{11}, X_{10}, and X_{01} are independently distributed Poisson variables with means v_{12}, (v_1-v_{12}), and (v_2-v_{12}) respectively (Holgate, 1964).

The marginal distributions of R_1 and R_2 are Poisson distributions with means v_1 and v_2 respectively, since

$$G_1(s_1) = G_{12}(s_1, 1) = \exp[v_1(s_1 - 1)],$$

with mean and variance:

$$E(r_1) = \text{var}(r_1) = v_1,$$

and

$$G_2(s_2) = G_{12}(1, s_2) = \exp[v_2(s_2 - 1)],$$

with mean and variance:

$$E(r_2) = \text{var}(r_2) = v_2.$$

Since the parameter v_{12} is the covariance of the BCP distribution, it follows that two uncorrelated Poisson variables distributed according to equation (8.3) are independent. The p.g.f. in such a case reduces to

$$G_{12}(s_1, s_2) = \exp[v_1(s_1 - 1) + v_2(s_2 - 1)],$$

and

$$P(r_1, r_2) = P(r_1)P(r_2) = \exp(-v_1)\frac{v_1^{r_1}}{r_1!}\exp(-v_2)\frac{v_2^{r_2}}{r_2!}$$

$$= \exp[-(v_1 + v_2)]\frac{v_1^{r_1} v_2^{r_2}}{r_1! r_2!}.$$

The BCP distribution describes the spatial dispersion of two classes of randomly distributed points. The marginal means of this distribution are v_1 and v_2 respectively, and v_{12} is the covariance. Note that, because the parameter space is subject to the inequality

$$0 \leq v_{12} \leq \min(v_1, v_2),$$

the correlation between the number of type 1 and type 2 points in a cell,

$$\rho = \frac{v_{12}}{(v_1 v_2)^{1/2}}, \tag{8.7}$$

is restricted to values between 0 and $\min[(v_1/v_2)^{1/2}, (v_2/v_1)^{1/2}]$. This can be shown by first supposing that $\min(v_1, v_2)$ is v_1, from which it follows that

$$0 \leq v_{12} \leq v_1,$$

and

$$0 \leq \frac{v_{12}^2}{v_1 v_2} \leq \frac{v_1}{v_2},$$

or

$$0 \leq \frac{v_{12}}{(v_1 v_2)^{1/2}} \leq \left(\frac{v_1}{v_2}\right)^{1/2}.$$

Similarly, if $\min(v_1, v_2)$ is v_2, then

$$0 \leq \frac{v_{12}}{(v_1 v_2)^{\frac{1}{2}}} \leq \left(\frac{v_2}{v_1}\right)^{\frac{1}{2}},$$

hence

$$0 \leq \rho \leq \min\left[\left(\frac{v_1}{v_2}\right)^{\frac{1}{2}}, \left(\frac{v_2}{v_1}\right)^{\frac{1}{2}}\right]. \tag{8.8}$$

8.3 Fundamental component distributions: a bivariate correlated negative binomial model

Recall the methods used in chapter 2 to derive the fundamental component distributions, and consider the following generalization of the procedure used there to obtain the univariate negative binomial distribution. Imagine a study region that has been gridded into square cells of a given dimension. Assume that initially (that is, at time $t = 0$) none of the cells contains any points. Let $p(r_1, t_1)$ be the probability that an individual grid cell has r_1 points of type 1 by time t_1. Assume that during the time interval $(t_1, t_1 + dt_1)$ a point locates in a particular cell, which already has r_1 type 1 points, with probability $f(r_1, t_1) dt_1$, and that the time interval is so short that no more than one point can locate in a particular cell during a single time interval. It then follows, by the arguments presented in chapter 2, that

$$\frac{\partial}{\partial t_1} G_1(s_1; t_1) = (s_1 - 1) L_1(s_1; t_1),$$

where $G_1(s_1; t_1)$ is the probability generating function of the random variable R_1 [see equation (2.2)], and

$$L_1(s_1; t_1) = \sum_{r_1 = 0}^{\infty} f(r_1, t_1) p(r_1, t_1) s^{r_1}.$$

Now assume that the probability that a type 1 point locates in a cell increases linearly with the number of such points already in that cell, as follows:

$$f(r_1, t_1) = c + b r_1, \qquad c > 0, b > 0.$$

Then

$$\frac{\partial}{\partial t_1} G_1(s_1; t_1) = (s_1 - 1)\left[c G_1(s_1; t_1) + b s_1 \frac{\partial}{\partial s_1} G_1(s_1; t_1)\right].$$

Whence

$$G_1(s_1; t_1) = \{\exp(b t_1) - [\exp(b t_1) - 1] s_1\}^{-c/b},$$

or

$$G_1(s_1) = (1 + p_1 - p_1 s_1)^{-k},$$

where $p_1 = \exp(b t_1) - 1$ and $k = c/b$.

Thus the probability that r_1 type 1 points occur in a cell is given by

$$P(r_1) = \binom{k+r_1-1}{r_1}\left(\frac{p_1}{1+p_1}\right)^{r_1}\left(\frac{1}{1+p_1}\right)^k.$$

Imagine now that another class of points, called type 2 points, locate in the same planar region in a manner similar to that followed by type 1 points, but with a probability that increases linearly with the sum of type 1 *and* type 2 points already in the cell, as follows:

$$f(r_2, t_2|r_1) = c + b(r_1+r_2), \quad c > 0, b > 0.$$

Then

$$G_2(s_2; t_2|r_1) = \{\exp(bt_2) - [\exp(bt_2) - 1]s_2\}^{-(c/b)-r_1},$$

or

$$G_2(s_2|r_1) = (1+p_2-p_2s_2)^{-(k+r_1)},$$

where $p_2 = \exp(bt_2) - 1$ and $k = c/b$, as before. Hence

$$P(r_2|r_1) = \binom{k+r_1+r_2-1}{r_2}\left(\frac{p_2}{1+p_2}\right)^{r_2}\left(\frac{1}{1+p_2}\right)^{k+r_1}.$$

Since

$$P(r_1, r_2) = P(r_2|r_1)P(r_1),$$

we have that

$$P(r_1, r_2) = \binom{k+r_1+r_2-1}{r_1 \ r_2}\left[\frac{p_1}{(1+p_1)(1+p_2)}\right]^{r_1}$$

$$\times \left(\frac{p_2}{1+p_2}\right)^{r_2}\left[\frac{1}{(1+p_1)(1+p_2)}\right]^k$$

$$= \binom{k+r_1+r_2-1}{r_1 \ r_2} Q_1^{r_1} Q_2^{r_2} P^k, \tag{8.9}$$

where $P = 1/(1+p_1)(1+p_2)$, $Q_1 = p_1/(1+p_1)(1+p_2)$, and $Q_2 = 1-P-Q_1$. Alternatively

$$P(r_1, r_2) = \binom{k+r_1+r_2-1}{r_1 \ r_2} p_1^{r_1}[(1+p_1)p_2]^{r_2} q^{-k-r_1-r_2}, \tag{8.10}$$

where $q = (1+p_1)(1+p_2) = 1+p_1+p_2+p_1p_2$.

Associated with these alternative expressions of the 'contagious' *bivariate correlated negative binomial* (BCNB) model are the following probability generating functions:

$$G_{12}(s_1, s_2) = \left(\frac{P}{1-Q_1s_1-Q_2s_2}\right)^k, \tag{8.11}$$

$$= (q - p_1s_1 - p_2s_2 - p_1p_2s_2)^{-k}. \tag{8.12}$$

The marginal distribution of R_1 is a univariate negative binomial distribution with p.g.f.

$$G_1(s_1) = G_{12}(s_1, 1) = (1 + p_1 - p_1 s_1)^{-k},$$

with parameters k, p_1, and mean and variance:

$$E(r_1) = kp_1,$$

$$\text{var}(r_1) = kp_1(1 + p_1).$$

The marginal distribution of R_2 is also a univariate negative binomial distribution. Its p.g.f. is

$$G_2(s_2) = G_{12}(1, s_2) = [1 + (1 + p_1)p_2 - (1 + p_1)p_2 s_2]^{-k}$$

with parameters k, p_1, p_2, and mean and variance:

$$E(r_2) = k(1 + p_1)p_2,$$

$$\text{var}(r_2) = k[(1 + p_1)p_2][1 + (1 + p_1)p_2].$$

The correlation between the number of type 1 and type 2 points in a cell is

$$\rho = \left[\left(1 + \frac{1}{p_1}\right)\left(1 + \frac{1}{(1 + p_1)p_2}\right)\right]^{-\frac{1}{2}}. \tag{8.13}$$

8.4 Bivariate compound distributions—another bivariate correlated negative binomial model

Recall the p.g.f. of the bivariate correlated Poisson distribution that was presented in equation (8.5), and let $v_1 = \lambda a_1$, $v_2 = \lambda a_2$, and $v_{12} = \lambda a_{12}$. Then

$$G_{12}(s_1, s_2 | \lambda) = \exp\{\lambda[(a_1 - a_{12})(s_1 - 1) + (a_2 - a_{12})(s_2 - 1) + a_{12}(s_1 s_2 - 1)]\}, \tag{8.14}$$

where $\lambda > 0$, and $0 \leq a_{12} \leq \min(a_1, a_2)$.

Expressed in this manner, the BCP distribution may be viewed as a distribution that is conditional on the parameter λ. Thus we have that

$$G_{12}(s_1, s_2) = \int_0^\infty G_{12}(s_1, s_2 | \lambda) P(\lambda) \, d\lambda \tag{8.15}$$

$$= \int_0^\infty \exp(\lambda \psi) P(\lambda) \, d\lambda$$

$$= M_\lambda(\psi), \tag{8.16}$$

where $M_\lambda(\psi)$ is the m.g.f. of $P(\lambda)$, and ψ represents the entire expression inside the square brackets in equation (8.14).

If we now compound the conditional bivariate Poisson distribution in equation (8.14) with a gamma distribution

$$P(\lambda) = \frac{x^k \lambda^{k-1}}{\Gamma(k)} \exp(-\lambda x), \qquad \lambda > 0, x > 0, k > 0,$$

and

$$M_\lambda(\theta) = \left(1 - \frac{\theta}{x}\right)^{-k},$$

then the p.g.f. of the resulting unconditional bivariate distribution is

$$\begin{aligned}
G_{12}(s_1, s_2) &= \left(1 - \frac{\psi}{x}\right)^{-k} \\
&= \left[1 - \frac{(a_1 - a_{12})}{x}(s_1 - 1) - \frac{(a_2 - a_{12})}{x}(s_2 - 1) - \frac{a_{12}}{x}(s_1 s_2 - 1)\right]^{-k} \\
&= [1 - p_1(s_1 - 1) - p_2(s_2 - 1) - p_{12}(s_1 s_2 - 1)]^{-k} \\
&= (q - p_1 s_1 - p_2 s_2 - p_{12} s_1 s_2)^{-k}, \qquad (8.17)
\end{aligned}$$

where

$$p_1 = \frac{a_1 - a_{12}}{x} > 0,$$

$$p_2 = \frac{a_2 - a_{12}}{x} > 0,$$

$$p_{12} = \frac{a_{12}}{x} > 0,$$

$$q = 1 + p_1 + p_2 + p_{12} > 0.$$

Edwards and Gurland (1961) refer to this distribution as a compound bivariate correlated Poisson distribution. We shall, however, include it in the class of bivariate correlated negative binomial distributions. In order to distinguish it from the 'contagious' BCNB distribution, derived in the preceding subsection, we shall henceforth call this distribution a BCNB2 distribution, and the 'contagious' distribution a BCNB1 distribution.

Several recurrence relationships can be established for obtaining the probabilities $P(r_1, r_2)$. Two suggested by Edwards and Gurland (1961) are:

$$P(r_1, r_2) = (-q)^{-r_1}(-p_1)^{r_1 - r_2}(-p_{12})^{r_2} \sum_{i=0}^{r_2} \left[\frac{\Gamma(k + r_1 + i)}{(r_1 - r_2 + i)!(r_2 - i)!\Gamma(k + i)}\right]$$
$$\times \left(\frac{p_1}{p_{12}}\right)^i P(0, i), \qquad (8.18)$$

$$P(r_1 + j, r_1) = \left(\frac{p_1}{p_2}\right)^j P(r_1, r_1 + j), \qquad (8.19)$$

where $r_1 \geq r_2$, and $r_1, r_2, i, j = 0, 1, 2, \ldots$, and the first term is
$P(0, 0) = G_{12}(0, 0) = q^{-k}$.

The marginal distribution of R_1 is a univariate negative binomial distribution:
$$G_1(s_1) = G_{12}(s_1, 1) = (q - p_1 s_1 - p_2 - p_{12} s_1)^{-k}$$
$$= [1 + (p_1 + p_{12}) - (p_1 + p_{12}) s_1]^{-k},$$
with parameters k, p_1, and p_{12}, and mean and variance:
$$E(r_1) = k(p_1 + p_{12}),$$
$$\text{var}(r_1) = k(p_1 + p_{12})(1 + p_1 + p_{12}).$$

Similarly the marginal distribution of R_2 is also a univariate negative binomial distribution with parameters k, p_2, and p_{12}, and mean and variance:
$$E(r_2) = k(p_2 + p_{12}),$$
$$\text{var}(r_2) = k(p_2 + p_{12})(1 + p_2 + p_{12}).$$

The correlation between the number of type 1 and type 2 points in a cell is
$$\rho = \frac{p_{12}(p_1 + p_2 + p_{12} + 1) + p_1 p_2}{[(p_1 + p_{12})(1 + p_1 + p_{12})(p_2 + p_{12})(1 + p_2 + p_{12})]^{\frac{1}{2}}}. \tag{8.20}$$

8.5 Bivariate generalized distributions—a bivariate correlated Neyman Type A model

As with compound distributions, it is a relatively simple matter to extend the notion of generalized distributions to the bivariate case. Several alternative formulations are possible, each leading to slightly different results. For example, we may postulate a clustering process that is univariate with respect to the distribution of *clusters* in space, but which assigns a bivariate distribution to the number of type 1 and type 2 *points* that occur in each cluster. Alternatively we may assume that type 1 and type 2 point clusters have a specified bivariate spatial distribution and that the number of type i points in a type i point cluster is defined by a univariate distribution. We shall focus on the latter model. Holgate (1966) examines both of these models and includes yet a third formulation.

It is easily established that the bivariate generalized distribution which we wish to study has a probability generating function of the form
$$G_{12}(s_1, s_2) = H_{12}[K_1(s_1), K_2(s_2)], \tag{8.21}$$
where $H_{12}[.], K_1(s_1)$, and $K_2(s_2)$ are respectively the p.g.f.'s of the bivariate distribution defining the number of type 1 and type 2 point

clusters in each cell, and the two univariate distributions defining the number of type 1 and type 2 *points* in each point cluster. Thus, for generalized bivariate correlated Poisson distributions, we have that

$$G_{12}(s_1,s_2) = \exp\{(v_1-v_{12})[K_1(s_1)-1]+(v_2-v_{12})[K_2(s_2)-1] \\ + v_{12}[K_1(s_1)K_2(s_2)-1]\}, \qquad (8.22)$$

and, assuming that the two univariate distributions are both Poisson distributions with means a_1 and a_2 respectively, we obtain

$$G_{12}(s_1,s_2) = \exp[\![(v_1-v_{12})\{\exp[a_1(s_1-1)]-1\} \\ + (v_2-v_{12})\{\exp[a_2(s_2-1)]-1\} \\ + v_{12}\{\exp[a_1(s_1-1)+a_2(s_2-1)]-1\}]\!]. \qquad (8.23)$$

This is the p.g.f. of a particular bivariate generalization of the univariate Neyman Type A distribution. Differentiation of equation (8.23) r_1+1 times with respect to v_1, and r_2 times with respect to v_2, leads, on setting $v_1 = v_2 = 0$, to the recurrence relationship (Holgate, 1966)

$$P(r_1+1, r_2) = \frac{v_{12}a_1\exp[-(a_1+a_2)]}{r_1+1} \sum_{i=0}^{r_1}\sum_{j=0}^{r_2} \frac{a_1^i a_2^j}{i!j!} P(r_1-i, r_2-j) \\ + \frac{(v_1-v_{12})a_1\exp(-a_1)}{r_1+1} \sum_{i=0}^{r_1} \frac{a_1^i}{i!} P(r_1-i, r_2), \qquad (8.24)$$

with

$$P(0,0) = G_{12}(0,0) = \exp[\![(v_1-v_{12})[\exp(-a_1)-1] \\ + (v_2-v_{12})[\exp(-a_2)-1] \\ + v_{12}\{\exp[-(a_1+a_2)]-1\}]\!],$$

where v_1 and v_2 are the marginal means of the bivariate correlated Poisson distribution, and v_{12} is the covariance

$$0 \leqslant v_{12} \leqslant \min(v_1, v_2).$$

The marginal distributions of the above derived *bivariate correlated Neyman Type A* (BCNTA) distribution are univariate Neyman Type A distributions:

$$G_1(s_1) = G_{12}(s_1, 1) = \exp[\![v_1\{\exp[a_1(s_1-1)]-1\}]\!]$$

with mean and variance:

$$E(r_1) = v_1 a_1,$$
$$\text{var}(r_1) = v_1 a_1(a_1+1);$$

and

$$G_2(s_2) = G_{12}(1, s_2) = \exp[\![v_2\{\exp[a_2(s_2-1)]-1\}]\!],$$

with mean and variance:

$E(r_2) = v_2 a_2$,

$\text{var}(r_2) = v_2 a_2 (a_2 + 1)$.

The covariance is

$\text{cov}(r_1, r_2) = v_{12} a_1 a_2$,

and the correlation coefficient between the number of type 1 and type 2 points in a cell is

$$\rho = \frac{v_{12} a_1 a_2}{[v_1 a_1 (a_1 + 1) v_2 a_2 (a_2 + 1)]^{1/2}} \cdot \quad (8.25)$$

8.6 Bivariate analysis of urban spatial dispersion in Ljubljana, Yugoslavia

We have seen in chapter 4 that, in order to carry out statistical tests on the goodness of fit of a hypothesized theoretical model to an observed empirical distribution, it is first necessary to obtain estimates of the theoretical model's parameters. The resulting fitted theoretical distribution may then be used to generate the expected frequency distribution that is to be compared with the observed one. Such a comparison commonly is effected through a chi-square goodness of fit test.

The same general procedure applies in the bivariate case. To illustrate this, the models described above have been fitted, using moment estimators, to the observed bivariate distribution of the number of food stores and people in our study area in Ljubljana, Yugoslavia. The observed bivariate frequency distribution, for our 10 by 10 grid system with quadrats that are 300 m on a side, appears in table 8.1. Table 8.2 presents the associated sample statistics, parameter estimates, and chi-square statistics. Tables 8.3, 8.4, 8.5, and 8.6 give the corresponding expected bivariate frequency distributions.

Moment estimates of v_1, v_2, and v_{12} in the BCP model were obtained by equating these three parameters to their corresponding sample statistics—the sample means of the two marginal distributions and the sample covariance—subject to the restriction that the covariance must not be negative and must not exceed the smaller of the two marginal sample means.

Five sample statistics (the two marginal sample means, the two marginal sample variances, and the sample covariance) may be used in various combinations to derive moment estimates of the three parameters in the BCNB1 model—k, p_1, and p_2. Thus $\binom{5}{3} - 2 = 8$ alternative sets of such moment estimates may be calculated. Rather than compute all of them, parameter estimates were instead obtained with a procedure, suggested by Arbous and Kerrich (1951), that uses the mean and variance of the sum of the two random variables and the quotient of their individual marginal sample means (Rogers and Martin, 1971).

Bivariate distributions

Table 8.1. Observed bivariate distribution of food stores and population in Ljubljana, Yugoslavia.

Population per cell r_1 (500's)	Food stores per cell r_2										Marginal cell total for r_1
	0	1	2	3	4	5	6	7	8	9+	
0	35	4	5	0	0	0	0	0	0	0	44
1	10	6	3	2	3	1	0	0	0	1	26
2	3	1	1	3	1	1	2	0	0	2	14
3	0	2	1	1	2	1	0	0	0	3	10
4	1	0	1	0	0	1	0	0	0	0	3
5	0	0	0	0	1	1	0	0	0	1	3
Marginal cell total for r_2	49	13	11	6	7	5	2	0	0	7	100

Table 8.2. Sample statistics, parameter estimates, and chi-square statistics.

1 *Sample statistics*

$\widehat{E(r_1)} = 1\cdot1100$ $\widehat{\text{var}(r_1)} = 1\cdot7353$ $\widehat{\text{cov}(r_1,r_2)} = 2\cdot4189$

$\widehat{E(r_2)} = 2\cdot0100$ $\widehat{\text{var}(r_2)} = 11\cdot5454$ $\widehat{\text{corr}(r_1,r_2)} = 0\cdot5404$

2 *Parameter estimates and chi-square statistics*

  a Bivariate correlated Poisson model   d.f. $P_{0\cdot05}$

     $\hat{v}_1 = 1\cdot1100$ Along rows[a]: 72·53   10   18·30
     $\hat{v}_2 = 2\cdot0100$ columns[a]: 84·04   10   18·30
     $\hat{v}_{12} = 1\cdot1100$ diagonals[a]: 89·96   9   16·90

  b Bivariate correlated negative binomial model: 1

     $\hat{k} = 0\cdot6469$ Along rows[a]: 26·44   16   26·30
     $\hat{p}_1 = 1\cdot7159$ columns[a]: 21·88   15   25·00
     $\hat{p}_2 = 1\cdot1441$ diagonals[a]: 29·71   15   25·00

  c Bivariate correlated negative binomial model: 2

     $\hat{k} = 0\cdot4237$ Along rows[a]: 31·17   13   22·40
     $\hat{p}_1 = 2\cdot6098$ columns: 14·97   13   22·40
     $\hat{p}_2 = 4\cdot7340$ diagonals: 15·64   12   21·00
     $\hat{p}_{12} = 0\cdot0100$[c]
     $\hat{q} = 8\cdot3538$

  d Bivariate correlated Neyman Type *A* model

     $\hat{v}_1 = 1\cdot9706$ Along rows[a]: 80·78   11   19·70
     $\hat{a}_1 = 0\cdot5633$ columns[a]: 144·83   13   22·40
     $\hat{v}_2 = 0\cdot4237$ diagonals: 19·75   12   21·00
     $\hat{a}_2 = 4\cdot7440$
     $\hat{v}_{12} = 0\cdot4237$[b]

[a] Null hypothesis rejected at the 5% level of significance.
[b] Since $0 \leq v_{12} \leq \min(v_1, v_2)$, our computer program sets v_{12} equal to $\min(\hat{v}_1, \hat{v}_2)$ whenever the estimation process produces a v_{12} that is greater than $\min(\hat{v}_1, \hat{v}_2)$.
[c] Since p_{12} must be positive, our computer program arbitrarily sets $\hat{p}_{12} = 0\cdot0100$ whenever the estimation process produces a $\hat{p}_{12} \leq 0$.

Table 8.3. Expected bivariate distribution of food stores and population in Ljubljana, Yugoslavia: bivariate correlated Poisson model.

Population per cell r_1 (500's)	Food stores per cell r_2										Marginal cell total for r_1
	0	1	2	3	4	5	6	7	8	9+	
0	13·27	12·07	5·49	1·67	0·38	0·07	0·01	0	0	0	32·96
1	0·13	14·71	13·33	6·06	1·84	0·42	0·08	0·01	0	0	36·58
2	0	0·15	8·16	7·36	3·34	1·01	0·23	0·04	0·01	0	20·30
3	0	0	0·08	3·02	2·71	1·23	0·37	0·08	0·02	0	7·51
4	0	0	0	0·03	0·84	0·75	0·34	0·10	0·02	0	2·08
5	0	0	0	0	0·01	0·19	0·17	0·07	0·02	0·11	0·57
Marginal cell total for r_2	13·40	26·93	27·07	18·13	9·11	3·66	1·19	0·32	0·07	0·11	100·00

Table 8.4. Expected bivariate distribution of food stores and population in Ljubljana, Yugoslavia: bivariate correlated negative binomial model 1.

Population per cell r_1 (500's)	Food stores per cell r_2										Marginal cell total for r_1
	0	1	2	3	4	5	6	7	8	9+	
0	31·99	11·04	4·85	2·28	1·11	0·55	0·28	0·14	0·07	0·04	52·36
1	6·10	5·36	3·78	2·45	1·52	0·92	0·54	0·32	0·18	0·10	21·28
2	1·48	2·09	2·03	1·68	1·27	0·90	0·61	0·40	0·26	0·16	10·88
3	0·38	0·75	0·93	0·93	0·83	0·67	0·52	0·38	0·27	0·19	5·85
4	0·10	0·26	0·39	0·46	0·47	0·43	0·37	0·30	0·23	0·17	3·17
5	0·03	0·09	0·15	0·21	0·24	0·24	0·23	0·21	0·17	4·89	6·45
Marginal cell total for r_2	40·08	19·58	12·14	8·01	5·43	3·71	2·55	1·74	1·19	5·56	100·00

Table 8.5. Expected bivariate distribution of food stores and population in Ljubljana, Yugoslavia: bivariate correlated negative binomial model 2.

Population per cell r_1 (500's)	Food stores per cell r_2										Marginal cell total for r_1
	0	1	2	3	4	5	6	7	8	9+	
0	40·68	9·77	3·94	1·80	0·87	0·44	0·22	0·12	0·06	0·03	57·94
1	5·38	4·37	3·00	1·94	1·22	0·75	0·45	0·27	0·16	0·10	17·64
2	1·20	1·65	1·61	1·35	1·03	0·75	0·53	0·36	0·24	0·16	8·88
3	0·30	0·59	0·74	0·76	0·69	0·58	0·47	0·36	0·26	0·19	4·95
4	0·08	0·20	0·31	0·38	0·40	0·39	0·34	0·29	0·23	0·18	2·82
5	0·02	0·07	0·13	0·18	0·21	0·23	0·22	0·21	0·18	6·31	7·76
Marginal cell total for r_2	47·67	16·65	9·73	6·41	4·44	3·14	2·24	1·60	1·15	6·97	100·00

Four parameters need to be estimated in the BCNB2 model—k, p_1, p_2, and p_{12}. Combining the five sample statistics of marginal means, variances, and the covariance, we may obtain $\binom{5}{4} - 1 = 4$ alternative sets of moment estimates. We computed all four sets of moment estimators and report only the results of the best fitting expected bivariate distribution.

The BCNTA model has five parameters—$v_1, a_1, v_2, a_2,$ and v_{12}. Moment estimates of these parameters may be obtained by setting them equal to their corresponding observed values in the sample. As in the BCP distribution, we restrict the estimate of v_{12} to the range between zero and $\min(v_1, v_2)$.

Since inefficient estimators have been used, we cannot assert that the associated X^2 statistics are distributed as the chi-square distribution. However, moment estimators lead to a stricter test of the hypothesis than more efficient estimators. Thus, in the few instances where the X^2 statistic is not significant at the 5% level, we may conclude that the associated bivariate distribution accounts satisfactorily for the data.

As is customary in practical applications of the chi-square test, we have grouped neighboring classes to ensure that the expected number of observations in any frequency class should not be too small—in this instance not less than 2. However, unlike the case with univariate models, the process of grouping in bivariate models has an important element of arbitrariness that may influence profoundly the character of the findings. This arbitrariness enters in the decision regarding the direction along which to group the observations. That is, should the grouping process operate across the rows of the expected bivariate distribution, or should it instead be carried out down the columns or along the diagonals? Table 8.2 reveals that this choice has a considerable impact on the size of the X^2 statistic, and consequently on the decision whether to reject or fail to reject the null hypothesis.

Table 8.6. Expected bivariate distribution of food stores and population in Ljubljana, Yugoslavia: bivariate correlated Neyman Type A model.

Population per cell r_1 (500's)	Food stores per cell r_2										Marginal cell total for r_1
	0	1	2	3	4	5	6	7	8	9+	
0	33·70	0·34	0·80	1·27	1·52	1·46	1·19	0·86	0·57	1·09	42·79
1	16·76	0·36	0·85	1·35	1·62	1·58	1·31	0·97	0·68	1·57	27·02
2	8·89	0·24	0·56	0·90	1·08	1·06	0·89	0·68	0·50	1·34	16·13
3	3·92	0·13	0·30	0·48	0·58	0·57	0·49	0·38	0·29	0·88	8·02
4	1·57	0·06	0·14	0·22	0·27	0·27	0·23	0·19	0·14	0·49	3·57
5	0·78	0·03	0·08	0·13	0·16	0·16	0·14	0·11	0·09	0·78	2·46
Marginal cell total for r_1	65·61	1·14	2·72	4·34	5·22	5·10	4·26	3·20	2·27	6·14	100·00

Of all the expected bivariate models that appear in this paper, only the multivariate negative binomial seems to provide a semblance of an adequate fit to the data. Specifically, the BCNB1 model appears to fit the data reasonably well. The BCNB2 model's relatively poorer fit is probably a consequence of the crude method of estimation.

A comparison of the moment and maximum likelihood fits of the univariate Neyman Type A model to the marginals of the observed

Table 8.7. A comparison of moment and maximum likelihood estimates: univariate negative binomial and Neyman Type A distributions.

Population:					
Number of units per cell (500's)	Number of cells observed	Expected frequency			
		N.B. (mom.) $\hat{w} = 1 \cdot 1100$ $\hat{k} = 1 \cdot 9706$	N.B. (m.l.e.) $\hat{w} = 1 \cdot 1100$ $\hat{k} = 1 \cdot 6597$	N.T.A. (mom.) $\hat{v} = 1 \cdot 9706$ $\hat{a} = 0 \cdot 5633$	N.T.A. (m.l.e.) $\hat{v} = 1 \cdot 7831$ $\hat{a} = 0 \cdot 6225$
0	44	41·46	42·74	42·80	43·77
1	26	29·44	28·43	27·05	26·07
2	14	15·75	15·15	16·16	15·88
3	10	7·51	7·41	8·04	8·06
4	3	3·36	3·46	3·59	3·69
5	3	2·47	2·81	2·36	2·54
Total no. of cells = 100 $X^2 = 1 \cdot 73$		1·31	1·11	0·90	
Total no. of units = 93 $P_{0 \cdot 05} = 7 \cdot 81$		7·81	7·81	7·81	

$\hat{m}_1 = 1 \cdot 1100$ $\hat{m}_2 = 1 \cdot 7353$

[a] Each unit represents 500 people.

Food stores:					
Number of stores per cell	Number of cells observed	Expected frequency			
		N.B. (mom.) $\hat{w} = 2 \cdot 0100$ $\hat{k} = 0 \cdot 4237$	N.B. (m.l.e.) $\hat{w} = 2 \cdot 0100$ $\hat{k} = 0 \cdot 4130$	N.T.A. (mom.) $\hat{v} = 0 \cdot 4237$ $\hat{a} = 4 \cdot 7440$	N.T.A. (m.l.e.) $\hat{v} = 0 \cdot 8214$ $\hat{a} = 2 \cdot 4469$
0	49	47·68	48·16	65·70	47·22
1	13	16·68	16·50	1·15	8·22
2	11	9·81	9·67	2·74	10·77
3	6	6·55	6·45	4·36	9·99
4	7	4·63	4·57	5·25	7·67
5	5	3·38	3·34	5·12	5·39
6	2	2·52	2·50	4·28	3·69
7	0	1·91	1·90	3·21	2·49
8	0	1·47	1·46	2·28	1·66
9+	7	5·36	5·46	5·90	2·91
Total no. of cells = 100 $X^2 = 11 \cdot 49$		11·55	118·47	9·10	
Total no. of units = 201 $P_{0 \cdot 05} = 14 \cdot 10$		14·10	14·10	14·10	

$\hat{m}_1 = 2 \cdot 0100$ $\hat{m}_2 = 11 \cdot 5454$

distribution (table 8.7) suggests that a maximum likelihood estimator in the bivariate model would produce a much closer fit of the BCNTA distribution to the data. However, such estimators are complex, and algorithms for obtaining estimates are difficult to devise. Consequently an effective compromise may be to use the univariate maximum likelihood parameter estimates and the observed covariance to generate the expected bivariate distribution (table 8.8)[22]. However, in the case of the BCNB model, such a procedure would produce two estimates of k. Perhaps in this case k should be estimated using the distribution of $R = R_1 + R_2$.

Another reason for the generally unsatisfactory fits of the above bivariate models to the data is the asymmetrical character of the observed bivariate distribution. It is not surprising therefore that symmetrical bivariate models, having marginal distributions drawn from the same family of distributions, do not provide a satisfactory fit to the asymmetrical observed bivariate distribution. This asymmetry also contributes to the large differences that arise from alternative grouping procedures.

Finally it should be noted that these results, as with univariate models, are dependent on the size of the quadrat used to sample the data and on the width of the class interval with which these data are classified. A smaller sized quadrat would produce a greater number of empty cells, and a wider class interval for the population variable would lead to a similar result. It should be possible to extend the notion of 'optimal quadrat size' to include, in an analogous manner, the concept of 'optimal interval width'. Since the two are interrelated, they must be considered simultaneously.

Table 8.8. Expected bivariate distribution of food stores and population in Ljubljana, Yugoslavia: bivariate correlated Neyman Type A model using marginally estimated maximum likelihood estimates.

Population per cell r_1 (500's)	Food stores per cell r_2										Marginal cell total for r_1
	0	1	2	3	4	5	6	7	8	9+	
0	29·26	2·73	3·47	3·04	2·13	1·31	0·77	0·45	0·26	0·15	43·58
1	10·09	2·64	3·44	3·14	2·35	1·60	1·05	0·68	0·43	0·27	25·69
2	4·88	1·57	2·10	2·01	1·61	1·19	0·85	0·59	0·40	0·26	15·45
3	1·94	0·76	1·04	1·03	0·88	0·69	0·53	0·39	0·28	0·19	7·71
4	0·70	0·32	0·45	0·47	0·41	0·35	0·28	0·21	0·16	0·11	3·46
5	0·24	0·12	0·18	0·19	0·18	0·15	0·13	0·10	0·08	2·74	4·11
Marginal cell total for r_2	47·11	8·15	10·67	9·88	7·56	5·29	3·60	2·42	1·60	3·72	100·00

[22] The chi-square statistics generated by this expected bivariate distribution are

Along rows   24·19   d.f. = 17 $P_{0·05} = 27·60$
columns   20·61   d.f. = 16 $P_{0·05} = 26·30$
diagonals   17·46   d.f. = 16 $P_{0·05} = 26·30$.

Spatial sampling

9.1 Introduction

One of the most critical decisions in the design of a sampling scheme is that of determining a procedure for selecting the sampling units. Specifically, given a population, or *universe*, of $\mathcal{N}$ identifiable sampling units, how should one select N of these units for a sample? Since the only necessary condition in probability sampling is that each of the sampling units in the universe have a positive and known probability of being selected, a multitude of selection procedures may be defined.

A sampling procedure that is frequently adopted is *stratified random sampling*. In this method the universe is first divided into *strata*, and sampling units are then selected at random within each stratum, independently of the selection of units within the other strata. Although stratified random samples do not possess the convenient statistical properties of simple random samples, they almost always result in estimates that are more precise. For it is well-known that allocating sampling units to strata in proportion to the number of elements that are in the strata in the universe almost always results in sample estimates that have a smaller variance than those of a simple random sample of the same size.

Occasionally it may be desirable to stratify the universe according to two different criteria and form a two-way cross-stratification scheme. The number of stratifications that can be imposed on a sample, however, is limited by its size, since the number of observations must at least be as large as the number of strata generated by the ordinary double stratification technique. This problem has led statisticians to focus on the development of sampling controls that produce results which are similar to those obtained by additional stratification, but without increasing the sample size [for example, Goodman and Kish (1950) and Jessen (1970)].

In this chapter we use Jessen's (1970) method of probability sampling with marginal constraints to introduce spatial controls in the sampling of spatially distributed universes in order to ensure a greater geographical spread and balance than could be expected from simple random sampling. We begin with a short exposition of Jessen's method and then use it to estimate the population, the number of food stores, and their total volume of sales in Ljubljana, Yugoslavia. Then we link Jessen's method with the results of our bivariate quadrat analysis to develop a procedure for spatial sampling in situations where the various cross-stratification control totals are not known prior to the sampling and must be estimated on the basis of past experience.

Spatial sampling

9.2 Probability sampling with marginal constraints: the Jessen method
9.2.1 Definitions and notation

Following Jessen (1970), imagine a sampling frame consisting of a population of elements arranged in the form of a matrix with I rows and J columns. Let Y_{ij} and X_{ij} be two characteristics of interest possessed by an element in the ith row and jth column. Assume that X_{ij} is positively correlated with Y_{ij} and that it therefore may be regarded as a measure of the 'size' of Y_{ij}. Generally X_{ij} is known before the sample is drawn, but Y_{ij} is observed only when sampled. Let

$$X = \sum_{i=1}^{I} \sum_{j=1}^{J} X_{ij}, \tag{9.1}$$

and

$$A_{ij} = \frac{X_{ij}}{X}. \tag{9.2}$$

Thus A_{ij} may be interpreted as the relative measure of the size of the ijth element. Consequently

$$\sum_{i=1}^{I} \sum_{j=1}^{J} A_{ij} = A = 1. \tag{9.3}$$

Jessen's method of probability sampling with marginal constraints deals with the problem of selecting a sample of N elements out of this universe in such a way that the following constraints are satisfied:

$$E(N_{ij}) = NA_{ij}, \tag{9.4}$$

$$|N_{ij} - NA_{ij}| < 1, \tag{9.5}$$

$$|N_{.j} - NA_{.j}| < 1, \tag{9.6}$$

and

$$|N_{i.} - NA_{i.}| < 1, \tag{9.7}$$

where N_{ij} is the number of ijth elements selected,

$$A_{.j} = \sum_{i=1}^{I} A_{ij},$$

$$A_{i.} = \sum_{j=1}^{J} A_{ij},$$

$$N_{.j} = \sum_{i=1}^{I} N_{ij},$$

and

$$N_{i.} = \sum_{j=1}^{J} N_{ij}.$$

If $NA_{ij} \leq 1$, then, by virtue of the constraint in equation (9.5), N_{ij} must be either 0 or 1. Consequently

$$E(N_{ij}) = [0 \times P(N_{ij} = 0)] + [1 \times P(N_{ij} = 1)] = P(N_{ij} = 1) ;$$

whence, by equation (9.4),

$$P(N_{ij} = 1) = NA_{ij} \leq 1 . \qquad (9.8)$$

However, if $NA_{ij} > 1$, then, on defining $\bar{\bar{N}}_{ij}$ to be equal to the integer part of NA_{ij}, we have by virtue of the constraint in equation (9.5) that

$$\begin{aligned} E(N_{ij}) &= [\bar{\bar{N}}_{ij} \times P(N_{ij} = \bar{\bar{N}}_{ij})] + [\bar{\bar{N}}_{ij} + 1) \times P(N_{ij} = \bar{\bar{N}}_{ij} + 1)] \\ &= \bar{\bar{N}}_{ij}[P(N_{ij} = \bar{\bar{N}}_{ij}) + P(N_{ij} = \bar{\bar{N}}_{ij} + 1)] + P(N_{ij} = \bar{\bar{N}}_{ij} + 1) \\ &= \bar{\bar{N}}_{ij} + P(N_{ij} = \bar{\bar{N}}_{ij} + 1) ; \end{aligned} \qquad (9.9)$$

whence, by equation (9.4),

$$P(N_{ij} = \bar{\bar{N}}_{ij} + 1) = NA_{ij} - \bar{\bar{N}}_{ij} ,$$

or, defining

$$\overline{NA}_{ij} = NA_{ij} - \bar{\bar{N}}_{ij} , \qquad (9.10)$$

where

$$\overline{N} = N - \bar{\bar{N}} ,$$

we have that

$$P(N_{ij} = \bar{\bar{N}}_{ij} + 1) = \overline{NA}_{ij} \leq 1 . \qquad (9.11)$$

Note that the expression in equation (9.11) is also valid for $NA_{ij} \leq 1$, because when $NA_{ij} < 1$, $\bar{\bar{N}}_{ij} = 0$, and equation (9.11) reduces to equation (9.8), and, when $NA_{ij} = 1$, $\bar{\bar{N}}_{ij} = 1$, $P(N_{ij} = 1 + 1) = 0$, whence $P(N_{ij} = 1) = 1$.

Let

$$\overline{N}_{ij} = N_{ij} - \bar{\bar{N}}_{ij} , \qquad (9.12)$$

then we may view Jessen's method as a procedure for selecting $\overline{N}_{ij}$ ijth elements in a way such that

$$E(\overline{N}_{ij}) = \overline{NA}_{ij} \leq 1 , \qquad (9.13)$$

$$|\overline{N}_{ij} - \overline{NA}_{ij}| < 1 , \qquad (9.14)$$

$$|\overline{N}_{.j} - \overline{NA}_{.j}| < 1 , \qquad (9.15)$$

$$|\overline{N}_{i.} - \overline{NA}_{i.}| < 1 , \qquad (9.16)$$

and selecting an additional $\bar{\bar{N}}_{ij}$ ijth elements with certainty.

A simple numerical example may be instructive at this point. Suppose we have a cross-stratified universe of 100 elements with which we have associated a 4 × 3 'size' matrix, $[X_{ij}]$ say, which is bordered by the row and column vectors of marginal totals, $[X_{.j}]$ and $[X_{i.}]$ say, and by the grand total X, then

$$[X] = \begin{bmatrix} [X_{ij}] & | & [X_{i.}] \\ \hline [X_{.j}] & | & X \end{bmatrix} = \begin{bmatrix} 35 & 4 & 5 & | & 44 \\ 10 & 6 & 10 & | & 26 \\ 3 & 1 & 10 & | & 14 \\ 1 & 2 & 13 & | & 16 \\ \hline 49 & 13 & 38 & | & 100 \end{bmatrix} . \qquad (9.17)$$

We wish to draw samples of size $N = 10$ from this universe such that the constraints in equations (9.4), (9.5), (9.6), and (9.7) are all met. The desired expected sample allocations will be

$$[NA] = \begin{bmatrix} [NA_{ij}] & | & [NA_{i.}] \\ \hline [NA_{.j}] & | & NA = N \end{bmatrix} = \begin{bmatrix} 3 \cdot 5 & 0 \cdot 4 & 0 \cdot 5 & | & 4 \cdot 4 \\ 1 \cdot 0 & 0 \cdot 6 & 1 \cdot 0 & | & 2 \cdot 6 \\ 0 \cdot 3 & 0 \cdot 1 & 1 \cdot 0 & | & 1 \cdot 4 \\ 0 \cdot 1 & 0 \cdot 2 & 1 \cdot 3 & | & 1 \cdot 6 \\ \hline 4 \cdot 9 & 1 \cdot 3 & 3 \cdot 8 & | & 10 \cdot 0 \end{bmatrix} . \qquad (9.18)$$

Analogous definitions lead to

$$[\bar{\bar{N}}] = \begin{bmatrix} [\bar{\bar{N}}_{ij}] & | & [\bar{\bar{N}}_{i.}] \\ \hline [\bar{\bar{N}}_{.j}] & | & \bar{\bar{N}} \end{bmatrix} = \begin{bmatrix} 3 & 0 & 0 & | & 3 \\ 1 & 0 & 1 & | & 2 \\ 0 & 0 & 1 & | & 1 \\ 0 & 0 & 1 & | & 1 \\ \hline 4 & 0 & 3 & | & 7 \end{bmatrix} , \qquad (9.19)$$

and

$$[\overline{NA}] = \begin{bmatrix} [\overline{NA}_{ij}] & | & [\overline{NA}_{i.}] \\ \hline [\overline{NA}_{.j}] & | & \overline{NA} = \bar{N} \end{bmatrix} = \begin{bmatrix} 0 \cdot 5 & 0 \cdot 4 & 0 \cdot 5 & | & 1 \cdot 4 \\ 0 & 0 \cdot 6 & 0 & | & 0 \cdot 6 \\ 0 \cdot 3 & 0 \cdot 1 & 0 & | & 0 \cdot 4 \\ 0 \cdot 1 & 0 \cdot 2 & 0 \cdot 3 & | & 0 \cdot 6 \\ \hline 0 \cdot 9 & 1 \cdot 3 & 0 \cdot 8 & | & 3 \cdot 0 \end{bmatrix} . \qquad (9.20)$$

Thus we have the following matrix equivalent of equation (9.10):

$$[NA] = [\bar{\bar{N}}] + [\overline{NA}] , \qquad (9.21)$$

with the interpretation that $\bar{\bar{N}} = 7$ of the $N = 10$ members of the sample will be drawn with certainty, according to the cross-stratification defined by the matrix $[\bar{\bar{N}}]$, while the remaining $\bar{N} = 3$ members will be selected probabilistically, according to the cross-stratification defined by the matrix $[\overline{NA}]$, such that the constraint in equation (9.13) is always satisfied.

More specifically, we seek samples of 10 elements each that will include, for example, at least 4 but not more than 5 elements from the first row, averaging precisely 4·4 elements in the process, of which 3·5 elements, on the average, will be included in the first row and first column stratum.

Jessen shows how a set of samples that meet the constraints of equations (9.4) through (9.7) may be generated by an algorithm that appropriately 'decrements' the NA_{ij} each time a sample is selected. The size of the decrement associated with the particular sample so generated is then defined to be the probability of selecting that sample. The marginal constraints are imposed on the selection process by appropriately adapting an earlier procedure of Jessen's (Jessen, 1969).

9.2.2 The Jessen algorithm

The first step in Jessen's algorithm is to define, in possibly an arbitrary manner, a 'feasible' sample—a sample that meets the size and marginal constraints of equations (9.14) through (9.16). This may be carried out by placing unities in the appropriate locations of a 'zero-one' matrix, $[\overline{U^h_{ij}}]$ say, that is of the same order as the matrix $[\overline{NA}_{ij}]$. Such a sample will meet all of the size constraints and will receive an appropriate probability of selection, D^h say, which is calculated by a sequential decrementing of the total residual probability of selection, B^h say. We begin the algorithm by setting $B^0 = 1$ and $D^0 = 0$, and continue generating feasible samples until $B^h = 0$. The final sample is chosen from these generated feasible samples using the D^h as the probabilities of selection.

Some additional bookkeeping operations are set up by Jessen in order to generate a set of feasible samples that contains as few samples as possible, for a given sequence of $[\overline{U^h_{ij}}]$ matrices. Initially each ijth element is given a measure of size $B^1_{ij} = \overline{NA}_{ij}$. This quantity is then decremented by the amount D^h each time the element appears in a feasible sample. Another variable, C^h_{ij} say, is also computed at every step along the way in order to identify the largest permissible value for D^h.

The entire Jessen algorithm consists of the following sequence of steps:

Step 1. Initialize the algorithm by setting

(a) $B^1_{ij} = \overline{NA}_{ij}$,

(b) $C^0_{ij} = B^1_{ij} = \overline{NA}_{ij}$,

(c) $B^0 = 1$, $D^0 = 0$, and $h = 1$.

Step 2. Generate the hth feasible sample

(a) Define a sample of $\overline{N}$ elements that satisfies the marginal constraints by setting $\overline{U^h_{ij}} = 1$ if the ijth element is in the sample and to 0 otherwise.

(b) Compute the elements of the matrix $[C_{ij}^h]$ as follows:

$$C_{ij}^h = \begin{cases} B_{ij}^h & \text{if } \overline{U}_{ij}^h = 1 \\ B^{h-1} - B_{ij}^h & \text{if } \overline{U}_{ij}^h = 0 \end{cases}$$

or, equivalently,

$$C_{ij}^h = (1 - \overline{U}_{ij}^h)(B^{h-1} - B_{ij}^h) + \overline{U}_{ij}^h B_{ij}^h.$$

(c) Set $D^h = \min(C_{ij}^h)$.
(d) Compute $B^h = B^{h-1} - D^h$.
(e) Compute the elements of the matrix $[B_{ij}^{h+1}]$, as follows:

$$B_{ij}^{h+1} = \begin{cases} B_{ij}^h - D^h & \text{if } \overline{U}_{ij}^h = 1 \\ B_{ij}^h & \text{if } \overline{U}_{ij}^h = 0 \end{cases}$$

or, equivalently,

$$B_{ij}^{h+1} = (1 - \overline{U}_{ij}^h)B_{ij}^h + \overline{U}_{ij}^h(B_{ij}^h - D^h).$$

Step 3. If $B^h = 0$, STOP. Otherwise, increase the subscript h by 1, and go to step 2.

Applying Jessen's algorithm to our numerical example, we obtain the tableau presented in table 9.1 below. Observe that five feasible sampling schemes have been generated, and note that the probabilities associated with these schemes are $0 \cdot 3, 0 \cdot 3, 0 \cdot 2, 0 \cdot 1$, and $0 \cdot 1$ respectively. We conclude therefore that we should draw our sample of 10 elements in a manner such that seven elements are chosen with certainty according to the cross-classification scheme defined by the matrix $[\overline{N}_{ij}]$ in equation (9.19), and the remaining three elements are chosen probabilistically according to one of the five cross-classification schemes defined by the matrices $[\overline{U}_{ij}^h]$ in table 9.1, the hth scheme having a probability D^h of being selected.

The Jessen algorithm can lead to a situation in which it is impossible to satisfy one of the prescribed marginal constraints. For example, in table 9.1, if we select for our first feasible sample the matrix

$$[\overline{U}_{ij}^1] = \begin{bmatrix} 1 & 0 & 1 \\ 0 & 1 & 0 \\ 0 & 0 & 0 \\ 0 & 0 & 0 \end{bmatrix},$$

we find that $D^1 = 0 \cdot 5$ and therefore that the total probability remaining in the first row of $[B_{ij}^2]$ is $0 \cdot 4$. Since $B^1 = 1 - D^1 = 0 \cdot 5$, we see that it now becomes impossible for us to choose at least one element in this row in every generated sampling scheme, because to do this would require that the probability remaining in the first row be $0 \cdot 5$ and not $0 \cdot 4$.

This problem may be eliminated, however, by a slight modification of the Jessen algorithm in which steps 2(b) and 2(c) of the algorithm are expanded to include a consideration of the marginal row and column sums of the matrix $[C_{ij}^h]$.

Table 9.1. Tableau for selecting feasible samples from a 4 × 3 universe using Jessen's algorithm.

Sample h	$[\overline{U}^h_{ij}]$	$[C^h_{ij}]$	D^h	$[B^{h+1}_{ij}]$	B^h
0	$\begin{bmatrix} 0 & 0 & 0 \\ 0 & 0 & 0 \\ 0 & 0 & 0 \\ 0 & 0 & 0 \end{bmatrix}$	$\begin{bmatrix} 0\cdot5 & 0\cdot4 & 0\cdot5 \\ 0 & 0\cdot6 & 0 \\ 0\cdot3 & 0\cdot1 & 0 \\ 0\cdot1 & 0\cdot2 & 0\cdot3 \end{bmatrix}$	$0\cdot0$	$\begin{bmatrix} 0\cdot5 & 0\cdot4 & 0\cdot5 \\ 0 & 0\cdot6 & 0 \\ 0\cdot3 & 0\cdot1 & 0 \\ 0\cdot1 & 0\cdot2 & 0\cdot3 \end{bmatrix}$	$1\cdot0$
1	$\begin{bmatrix} 1 & 0 & 0 \\ 0 & 1 & 0 \\ 0 & 0 & 0 \\ 0 & 0 & 1 \end{bmatrix}$	$\begin{bmatrix} 0\cdot5 & 0\cdot6 & 0\cdot5 \\ 1\cdot0 & 0\cdot6 & 1\cdot0 \\ 0\cdot7 & 0\cdot9 & 1\cdot0 \\ 0\cdot9 & 0\cdot8 & 0\cdot3 \end{bmatrix}$	$0\cdot3$	$\begin{bmatrix} 0\cdot2 & 0\cdot4 & 0\cdot5 \\ 0 & 0\cdot3 & 0 \\ 0\cdot3 & 0\cdot1 & 0 \\ 0\cdot1 & 0\cdot2 & 0 \end{bmatrix}$	$0\cdot7$
2	$\begin{bmatrix} 0 & 1 & 1 \\ 0 & 0 & 0 \\ 1 & 0 & 0 \\ 0 & 0 & 0 \end{bmatrix}$	$\begin{bmatrix} 0\cdot5 & 0\cdot4 & 0\cdot5 \\ 0\cdot7 & 0\cdot4 & 0\cdot7 \\ 0\cdot3 & 0\cdot6 & 0\cdot7 \\ 0\cdot6 & 0\cdot5 & 0\cdot7 \end{bmatrix}$	$0\cdot3$	$\begin{bmatrix} 0\cdot2 & 0\cdot1 & 0\cdot2 \\ 0 & 0\cdot3 & 0 \\ 0 & 0\cdot1 & 0 \\ 0\cdot1 & 0\cdot2 & 0 \end{bmatrix}$	$0\cdot4$
3	$\begin{bmatrix} 1 & 0 & 0 \\ 0 & 1 & 0 \\ 0 & 0 & 0 \\ 0 & 1 & 0 \end{bmatrix}$	$\begin{bmatrix} 0\cdot2 & 0\cdot3 & 0\cdot2 \\ 0\cdot4 & 0\cdot3 & 0\cdot4 \\ 0\cdot4 & 0\cdot3 & 0\cdot4 \\ 0\cdot3 & 0\cdot2 & 0\cdot4 \end{bmatrix}$	$0\cdot2$	$\begin{bmatrix} 0 & 0\cdot1 & 0\cdot2 \\ 0 & 0\cdot1 & 0 \\ 0 & 0\cdot1 & 0 \\ 0\cdot1 & 0 & 0 \end{bmatrix}$	$0\cdot2$
4	$\begin{bmatrix} 0 & 1 & 1 \\ 0 & 0 & 0 \\ 0 & 0 & 0 \\ 1 & 0 & 0 \end{bmatrix}$	$\begin{bmatrix} 0\cdot2 & 0\cdot1 & 0\cdot2 \\ 0\cdot2 & 0\cdot1 & 0\cdot2 \\ 0\cdot2 & 0\cdot1 & 0\cdot2 \\ 0\cdot1 & 0\cdot2 & 0\cdot2 \end{bmatrix}$	$0\cdot1$	$\begin{bmatrix} 0 & 0 & 0\cdot1 \\ 0 & 0\cdot1 & 0 \\ 0 & 0\cdot1 & 0 \\ 0 & 0 & 0 \end{bmatrix}$	$0\cdot1$
5	$\begin{bmatrix} 0 & 0 & 1 \\ 0 & 1 & 0 \\ 0 & 1 & 0 \\ 0 & 0 & 0 \end{bmatrix}$	$\begin{bmatrix} 0\cdot1 & 0\cdot1 & 0\cdot1 \\ 0\cdot1 & 0\cdot1 & 0\cdot1 \\ 0\cdot1 & 0\cdot1 & 0\cdot1 \\ 0\cdot1 & 0\cdot1 & 0\cdot1 \end{bmatrix}$	$0\cdot1$	$\begin{bmatrix} 0 & 0 & 0 \\ 0 & 0 & 0 \\ 0 & 0 & 0 \\ 0 & 0 & 0 \end{bmatrix}$	$0\cdot0$
		Total	$1\cdot0$		

9.3 Probability sampling with marginal constraints: quadrat spatial sampling

We define *spatial sampling* to be the process of drawing samples from a study area for which we have available some information regarding the geographical arrangement of the members of the universe being sampled. It seems reasonable to assume that quadrat analysis of the spatial dispersion of such elements could provide useful information for spatial sampling efforts. In what follows we shall illustrate this point by linking our discussion of bivariate quadrat analysis, in chapter 8 of this book, to the Jessen method of probability sampling with marginal constraints. We shall call this particular blend of quadrat analysis and constrained probability sampling, *quadrat spatial sampling*.

9.3.1 A procedure for quadrat spatial sampling

Imagine a study area that has been gridded into square cells of equal size called quadrats. Assume that the number of two different categories of establishments, type 1 and type 2 establishments say, in each quadrat has been noted and that the results have been collected and arranged in a bivariate array, as in table 9.2. Further let the entries in this array, X_{ij} say, be positively correlated with some other characteristic of interest, denoted by Y_{ij} say, so that they may be regarded as a measure of the 'size' of Y_{ij}. We then have a 'size' matrix $[X]$, such as is illustrated in equation (9.17), and may proceed to apply Jessen's method of probability sampling with marginal constraints to draw samples from this universe of quadrats in order to obtain estimates of, for example,

$$Y = \sum_{i=0}^{I} \sum_{j=0}^{J} Y_{ij} . \tag{9.22}$$

Note that the X_{ij} are presumed to be known for all i and j, but Y_{ij} is observed only when selected in a sample.

Recall the size matrix $[X]$ presented in equation (9.17), and consider its interpretation in the context of our current discussion. Of the 100 quadrats covering a given study area, 35 are totally empty, 4 contain no type 1 establishments but have exactly one type 2 establishment, 10 contain no type 2 establishments but have exactly one type 1 establishment, 6 contain exactly one type 1 establishment and one type 2 establishment, and so on. In order to draw a sample of $N = 10$ quadrats from this cross-stratified universe, such that the constraints in equations (9.4), (9.5), (9.6), and (9.7) are all met, we need to allocate our sample according to the expectations defined by the matrix $[NA]$ in equation (9.18). Thus, for example, we would always include in our sample 3 or 4 of the 35 empty quadrats and precisely 1 of the 10 quadrats that contain a single type 1 establishment but no type 2 establishments.

A number of alternative strategies may be adopted in selecting particular quadrats *from among those contained in each cross-classification*. Initially,

Table 9.2. Bivariate distribution of type 1 and type 2 establishments in a study area.

Number of type 1 establishments per quadrat	Number of type 2 establishments per quadrat						Marginal total
	0	1	2	.	.	.	
0	X_{00}	X_{01}	X_{02}	.	.	.	$X_{0\cdot}$
1	X_{10}	X_{11}	X_{12}	.	.	.	$X_{1\cdot}$
2	X_{20}	X_{21}	X_{22}	.	.	.	$X_{2\cdot}$
.	.	.	.				.
.	.	.	.				.
Marginal total	$X_{\cdot 0}$	$X_{\cdot 1}$	$X_{\cdot 2}$	.	.	.	X

we adopt the simplest procedure, which is to select these quadrats at random. Subsequently we modify this strategy by selecting quadrats according to probabilities that vary inversely with the quadrat's distance from the 'center' of the study area.

9.3.2 Quadrat spatial sampling of the retail and population distributions in Ljubljana, Yugoslavia

The effectiveness of the quadrat spatial sampling procedure outlined above has been tested with data describing the spatial dispersion of retail establishments and population in Ljubljana, Yugoslavia. As in the rest of this book our study area is the central region of Ljubljana which is situated inside the 3000 m by 3000 m grid that is illustrated in figure 7.1 of chapter 7.

In 1966 this study area contained 76575 people and 201 food stores, which together accounted for a total annual volume of sales of 301490000 dinars. We have elected to view the study area as a 10 by 10 grid of 100 quadrats, 300 m on a side. Thus each quadrat will, on the average, contain about 766 people and two food stores with a total annual volume of sales of approximately 3015000 dinars. These averages, however, may be somewhat deceiving because the spatial distribution of population and retail establishments is a clustered one, with many quadrats being empty and only a few having high concentrations of people and food stores. The exact bivariate distribution of population and food stores appeared in table 8.1 of chapter 8. It appears once again, in consolidated form, as the 'size' matrix $[X]$ in equation (9.17). Recall that the population counts are reported in units of 500 people each. Thus, for example, 44 of the 100 quadrats have less than 500 people in them, and 16 have a total residential population of 1500 or more.

Our vehicle for assessing the effectiveness of quadrat spatial sampling is a simulation computer program that draws S samples of size N, from the universe of 100 quadrats covering the Ljubljana study area, in accordance with the Jessen procedure described earlier and the observed bivariate distribution defined by the matrix $[X]$ in equation (9.17). For each of the S samples it computes the sample mean and multiplies it by 100 to obtain an estimate of the study area's total for the following three variables: total residential population, total number of food stores, and total volume of food store sales. The program then computes the means of these three sets of S totals and the associated standard deviations. Finally, using these means and standard deviations, it calculates two statistics that measure the relative precision of the quadrat spatial sampling procedure[23]: (1) the ratio of the estimated to the known mean, a measure we shall call the *bias ratio*;

---

[23] One sampling scheme will be said to be relatively more precise than another if the variance or mean square error of the estimate that is associated with it is less than that of the other. If the cost of both sampling schemes is approximately the same, then the relatively more precise sampling scheme also is the more *efficient* sampling scheme.

(2) the ratio of the estimated standard deviation to the known mean, an index we shall refer to as the *coefficient of variation*. Relatively *precise* sampling strategies are those that lead to a bias ratio that is close to unity and to a small coefficient of variation.

The results of the quadrat spatial sampling simulation, carried out with the Ljubljana data, are summarized in table 9.3. We include the results both for 10 and 20% samples (that is, $N = 10$ and $N = 20$) that were generated with simulations using 100 samplings (that is, $S = 100$). [Experiments with other sampling sizes ($S = 50$ and $S = 200$) indicated that the findings do not change appreciably with an increase in the number of simulated samplings.] Results obtained with simple random sampling are also included as a benchmark against which to evaluate the effectiveness of the stratifications imposed by quadrat spatial sampling.

The first conclusion that may be drawn from an examination of table 9.3 is that quadrat spatial sampling (qss) yields estimates that are relatively more precise than those obtained with simple random sampling (srs). In all cases reported in table 9.3, the coefficient of variation associated with the qss estimates is smaller than the corresponding coefficient for the srs estimates. Since the bias ratios of the qss estimates are always closer to unity than those of the srs estimates, we conclude that quadrat spatial sampling produces more precise estimates than simple random sampling.

The second conclusion that is indicated by table 9.3 is that the relative precision of quadrat spatial sampling over simple random sampling decreases as the spatial clustering of the elements of the universe being sampled increases. This is reflected by the dramatic differences in the degrees of improvement afforded by qss estimators over srs estimators for the three

Table 9.3. The relative precision of quadrat spatial sampling.

Variable	Sample size (%)	Bias ratio and coefficient of variation (in parentheses) for simple random sampling and quadrat spatial sampling	
		srs	qss
Population	10	1·039 (0·276)	0·991 (0·097)
	20	1·013 (0·174)	1·000 (0·062)
Number of food stores	10	1·119 (0·538)	0·972 (0·361)
	20	1·077 (0·360)	0·995 (0·239)
Volume of food store sales	10	1·026 (**0·901**)	0·978 (0·863)
	20	1·073 (0·697)	1·000 (0·576)

variables being sampled: population, number of food stores, and their total volume of sales. Whereas qss estimation of total population yields a coefficient of variation which is 2/3 smaller than that obtained by srs estimation, the corresponding results for total number of food stores and their volume of sales are progressively less favorable—a reduction of about 1/3 for the former and between 1/20 to 1/6 for the latter. Since we established in chapter 7 that food store sales volumes are more clustered spatially than food stores, which in turn are more clustered than the residential population that they serve, we conclude that the effectiveness of quadrat spatial sampling decreases with increasing spatial clustering.

Finally, the third conclusion that is suggested by table 9.3 is that a doubling of the sample size in quadrat spatial sampling increases precision by about a third. Moreover, although it is not surprising to find that a 20% sample always yields more precise estimates than a 10% one, a relatively unexpected result is that for sampling total population a 10% quadrat spatial sample is apparently about twice as precise as a 20% simple random sample. Note that the situation is reversed in the case of volume of sales sampling, and that both are about equally precise in the case of food store sampling.

9.3.3 Sensitivity tests of the quadrat spatial sampling procedure

Several of the steps that led to the generation of our particular sampling scheme were somewhat arbitrary; therefore it is appropriate for us to investigate the sensitivity of our findings to minor variations in these steps. Consequently we shall now briefly examine the sensitivity of the quadrat spatial sampling procedure to changes in the dimensions of the size matrix, to differences between the sets of feasible samples generated by the Jessen algorithm, and to a spatially biased selection of quadrats from among those included in a particular cross-classified stratum.

Table 9.4 presents the bias ratios and coefficients of variation for quadrat spatial sampling carried out with size matrices of different dimensions. (All of the results in table 9.4 and in subsequent tables were generated by simulations carried out with $S = 100$ samplings.) For simplicity only the number of food store strata is varied; thus only the number of columns in the matrix $[X_{ij}]$ is changed. The results indicate that an increase in the dimensions of the size matrix usually leads to an increase in efficiency. However, the degree of improvement is minimal. Hence we conclude that our quadrat spatial sampling procedure is relatively insensitive to differences in the order of the size matrix.

The Jessen algorithm does not generate a unique sampling scheme for a given size matrix. Thus it is important to determine whether different feasible sampling schemes are likely to produce different results. Table 9.5 presents some evidence to the contrary. In it we report results obtained with two different sets of feasible samples, scheme I and scheme II say, that were generated by applying the Jessen algorithm to the 4 by 5 size matrix for the

Spatial sampling

Table 9.4. Sensitivity of quadrat spatial sampling to differences in the dimensions of the size matrix.

Variable	Sample size (%)	Bias ratio and coefficient of variation (in parentheses) for alternative dimensions of the size matrix		
		4 × 1	4 × 3	4 × 5
Population	10	1·001 (0·148)	0·991 (0·097)	1·006 (0·094)
	20	0·995 (0·075)	1·000 (0·062)	1·006 (0·059)
Number of food stores	10	1·002 (0·442)	0·972 (0·361)	1·015 (0·325)
	20	0·992 (0·243)	0·995 (0·239)	1·008 (0·198)
Volume of food store sales	10	0·997 (0·907)	0·968 (0·863)	1·019 (0·819)
	20	1·018 (0·592)	1·000 (0·576)	1·030 (0·572)

Table 9.5. Sensitivity of quadrat spatial sampling to alternative feasible sampling schemes I and II.

Variable	Sample size (%)	Bias ratio and coefficient of variation (in parentheses) for alternative feasible sampling schemes[a]		Absolute value of the difference between I and II
		I	II	
Population	10	1·006 (0·094)	1·012 (0·114)	0·006 (0·020)
	20	1·006 (0·059)	0·995 (0·059)	0·011 (0·000)
Number of food stores	10	1·015 (0·325)	1·016 (0·328)	0·001 (0·003)
	20	1·008 (0·198)	0·988 (0·210)	0·020 (0·012)
Volume of food store sales	10	1·019 (0·819)	1·010 (0·822)	0·009 (0·003)
	20	1·030 (0·572)	0·979 (0·568)	0·051 (0·004)

[a] These results were generated with a 4 × 5 size matrix.

Ljubljana study area[24]. These results suggest the conclusion that the results of quadrat spatial sampling are relatively insensitive to differences among alternative feasible sampling schemes generated by the Jessen algorithm.

Finally, our procedure for selecting quadrats within each of the different strata has been to choose them at random. An obvious alternative is to select them according to some other decision rule. Table 9.6 compares results obtained by quadrat spatial sampling with random selection within

Table 9.6. Sensitivity of quadrat spatial sampling to spatially biased sample selection within strata.

Variable	Sample size (%)	Sampling scheme	Bias ratio and coefficient of variation (in parentheses) for random and spatially biased sample selection within strata[a]	
			random	biased
Population	10	I	1·006 (0·094)	1·022 (0·308)
		II	1·012 (0·114)	1·015 (0·291)
	20	I	1·006 (0·059)	1·015 (0·173)
		II	0·995 (0·059)	1·018 (0·176)
Number of food stores	10	I	1·015 (0·325)	1·023 (0·312)
		II	1·016 (0·328)	0·999 (0·314)
	20	I	1·008 (0·198)	1·017 (0·190)
		II	0·988 (0·210)	1·000 (0·203)
Volume of food store sales	10	I	1·019 (0·819)	0·994 (0·629)
		II	1·010 (0·822)	0·994 (0·644)
	20	I	1·030 (0·572)	1·031 (0·429)
		II	0·979 (0·568)	0·993 (0·426)

[a] These results were generated with a 4 × 5 size matrix.

[24] Although a 4 × 3 matrix is a convenient size for illustrating the Jessen algorithm, a 4 × 5 matrix provides us with a greater opportunity for generating substantially different sets of feasible samples. Moreover, as we have seen in table 9.4, the latter also leads to more precise qss estimates. Consequently we shall henceforth deal only with 4 × 5 size matrices.

strata to those obtained by the same sampling procedure but with
'spatially biased' selection within strata. (An alternative selection rule
would be to choose quadrats according to probabilities that are proportional
to their 'size'.) By spatially biased selection we mean the use of a decision
rule that assigns to each quadrat within a particular stratum a probability
of selection that varies inversely with that quadrat's distance from the
'center' of the study area, where the center of the study area is defined to
be the quadrat with the highest volume of total retail sales. Since the
probability of selecting a particular quadrat now varies with its location,
an appropriate weighting scheme must be introduced in the expansion of
sample totals to estimates of universe totals[25].

The results presented in table 9.6 show that spatially biased quadrat
spatial sampling reduced only the coefficients of variation associated with
the estimates of total volume of food store sales. And even here the
increase in precision did not come cheaply, for associated with this increase
was an increase in the **average** number of stores sampled[26]. The net
result is that a 20% increase in precision was obtained at approximately a
20% increase in sampling cost. Thus we conclude that the additional
control provided by a spatially biased selection rule is effective only when
the elements being sampled are highly concentrated spatially and when the
increase in precision that is thereby achieved is sufficiently important to
warrant the associated increase in sampling cost. In all other situations
such a selection rule does not appear to be effective and indeed may, as
in the case of total population, lead to a significant decrease in precision.

9.4 Quadrat spatial sampling with estimated constraints

Heretofore we have assumed that the spatial distribution of the elements
being sampled is given, and consequently that the size matrix and the
location of each cross-stratified quadrat are known with certainty. In many
practical applications of quadrat spatial sampling, however, we may not be
so fortunate. For example, we may know the bivariate size matrix, but

[25] The weight associated with each observation, for a sample of size N, drawn from a universe of size $\mathcal{N}$, in simple random sampling and quadrat spatial sampling is $\mathcal{N}/N$. The weight accorded to each observation in a sample of the same size for spatially biased quadrat spatial sampling is

$$\frac{\mathcal{N}}{N} \frac{1}{X_{ij} P_{ij}(h)},$$

where

$$P_{ij}(h) = \frac{1/d_h}{\sum_h 1/d_h},$$

and d_h is the aerial distance between the hth quadrat and the center of the study area.

[26] The average numbers of stores sampled in the 10% sample are approximately 22, 20, and 24 for the srs, qss, and spatially biased qss estimators respectively. For the 20% sample the corresponding totals are approximately 44, 40, and 48 respectively.

not the location of the quadrats. Or, occasionally, we may be ignorant of both. Therefore we shall now examine the effectiveness of quadrat spatial sampling in applications for which no current data are available. However, we will assume that a complete data census is available for some recent past period. Thus our problem reduces to one of using historical data to guide us in the design of a current quadrat spatial sampling scheme.

Suppose that through the use of historical data we have been able to estimate the current size matrix $[X_{ij}]$ for our study area. Assume that this matrix was obtained through the use of bivariate quadrat analysis methods described in chapter 8 of this book. For example, suppose that this matrix is the bivariate correlated Neyman Type A distribution presented in table 8.8. Rounding to the nearest integer, and consolidating that matrix to the dimensions 4×3, we obtain [27]

$$[\hat{X}] = \left[\begin{array}{c|c} [\hat{X}_{ij}] & [\hat{X}_{i.}] \\ \hline [\hat{X}_{.j}] & X \end{array}\right] = \begin{bmatrix} 29 & 3 & 12 & | & 44 \\ 10 & 3 & 13 & | & 26 \\ 5 & 1 & 9 & | & 15 \\ 3 & 1 & 11 & | & 15 \\ \hline 47 & 8 & 45 & | & 100 \end{bmatrix}. \qquad (9.23)$$

Since we do not have data for two time periods, we cannot carry out an actual test of the effectiveness of the estimated constraints. What we can do, however, is to compare the results generated by our 4×3 size matrix in equation (9.17) with those generated by the 4×3 size matrix in equation (9.23). This still leaves us with the problem of determining the spatial location of each cross-classified quadrat. Here we have no other recourse but to use the spatial locations actually observed at a previous time period. Thus our general approach consists of estimating the size matrix with a bivariate quadrat model and the spatial location of quadrats with the observed spatial distribution of a previous time period. Both estimates will lead to biases that increase with the degree to which the spatial structure of the current time period differs from that of the previous one.

Occasionally we encounter situations in which a previously empty stratum receives a positive entry in the estimated size matrix. We then cannot use the past census to estimate the location of quadrats in that particular stratum, and must either consolidate the size matrix in a way that eliminates the empty stratum, or artificially keep that stratum empty and 'adjust' the size matrix by some iterative technique such as the RAS method (Bacharach, 1970). For example, the observed 4×5 bivariate

---

[27] Compare this estimated size matrix with the observed one presented in equation (9.17).

size matrix for our study area contains two empty strata:

$$[X] = \begin{bmatrix} 35 & 4 & 5 & 0 & 0 & | & 44 \\ 10 & 6 & 3 & 2 & 5 & | & 26 \\ 3 & 1 & 1 & 3 & 6 & | & 14 \\ \underline{1} & \underline{2} & \underline{2} & \underline{1} & \underline{10} & | & \underline{16} \\ 49 & 13 & 11 & 6 & 21 & | & 100 \end{bmatrix}. \quad (9.24)$$

The associated estimated BCNTA distribution, however, does not:

$$[\hat{X}] = \begin{bmatrix} 29\cdot26 & 2\cdot73 & 3\cdot47 & 3\cdot04 & 5\cdot08 & | & 43\cdot58 \\ 10\cdot09 & 2\cdot64 & 3\cdot44 & 3\cdot14 & 6\cdot38 & | & 25\cdot69 \\ 4\cdot88 & 1\cdot57 & 2\cdot10 & 2\cdot01 & 4\cdot89 & | & 15\cdot45 \\ \underline{2\cdot88} & \underline{1\cdot21} & \underline{1\cdot66} & \underline{1\cdot69} & \underline{7\cdot84} & | & \underline{15\cdot28} \\ 47\cdot11 & 8\cdot15 & 10\cdot67 & 9\cdot88 & 24\cdot19 & | & 100\cdot00 \end{bmatrix}. \quad (9.25)$$

To resolve this problem, therefore, we may consolidate equation (9.25) to a 4 × 3 matrix and obtain equation (9.23), or we may set the appropriate two entries in equation (9.25) equal to zero and then adjust the elements of the size matrix, with the RAS method say, so that they sum to the marginal totals. Rounding the resulting matrix, we obtain

$$[\hat{X}]_{\text{adjusted}} = \begin{bmatrix} 35 & 4 & 5 & 0 & 0 & | & 44 \\ 7 & 2 & 3 & 5 & 9 & | & 26 \\ 3 & 1 & 2 & 3 & 6 & | & 15 \\ \underline{2} & \underline{1} & \underline{1} & \underline{2} & \underline{9} & | & \underline{15} \\ 47 & 8 & 11 & 10 & 24 & | & 100 \end{bmatrix}. \quad (9.26)$$

Table 9.7 presents the results that were obtained with these two alternative ways of resolving the 'empty stratum' problem. The previously generated results using the observed 4 × 3 and 4 × 5 size matrices also are included for comparison. The conclusion that such a comparison suggests is that the use of estimated controls in quadrat spatial sampling probably does not lead to serious losses in precision (but note the increases that occur in the bias ratios). However, this conclusion is necessarily a very tentative one and awaits further empirical support from studies that have spatial data for at least two different points in time.

Table 9.7. Quadrat spatial sampling with observed and estimated constraints.

Variable	Sample size (%)	Dimensions of the size matrix	Bias ratio and coefficient of variation (in parentheses) for qss estimation with observed and estimated constraints for 4 × 3 and 4 × 5 size matrices[a]	
			observed	estimated
Population	10	4 × 3	0·991 (0·097)	1·000 (0·103)
		4 × 5	1·006 (0·094)	0·986 (0·090)
	20	4 × 3	1·000 (0·062)	0·999 (0·062)
		4 × 5	1·006 (0·059)	0·987 (0·052)
Number of food stores	10	4 × 3	0·972 (0·361)	1·035 (0·344)
		4 × 5	1·015 (0·325)	1·051 (0·305)
	20	4 × 3	0·995 (0·239)	1·001 (0·239)
		4 × 5	1·008 (0·198)	1·107 (0·197)
Volume of food store sales	10	4 × 3	0·978 (0·863)	1·032 (0·870)
		4 × 5	1·019 (0·819)	0·915 (0·734)
	20	4 × 3	1·000 (0·576)	0·887 (0·514)
		4 × 5	1·030 (0·572)	1·074 (0·591)

9.5 Conclusion

Survey sampling is concerned with methods for selecting and analyzing a part of a universe in order to make inferences about the entire universe. Sampling offers the advantages of economy, speed, and, in certain circumstances, quality and accuracy.

Suppose that we wish to conduct a sample survey of food stores in a city such as Ljubljana, Yugoslavia, in order to estimate their average annual volume of sales. It is well-known that the majority of such stores are either small or middle-sized, and that there are only a few very large supermarkets. The latter, however, account for a considerable proportion of the total volume of sales. In other words, the size distribution of food stores is a highly skewed one, with a small proportion of stores accounting for a relatively large proportion of the total volume of sales. A simple random sample of these stores, therefore, could include either none or too many of these supermarkets and, in such instances, the sample would not adequately represent the universe. Thus a stratification of stores by size is commonly recommended for such sample surveys.

Now consider the same sampling problem from a spatial point of view. As we have seen, the spatial distribution of food stores is a clustered one, with a small proportion of quadrats accounting for a relatively large proportion of the total volume of sales. Consequently a simple spatial random sample may include either too few or too many of the stores in the 'downtown' district, and, in such instances the sample will not geographically represent the universe. Thus a stratification of stores by location seems to be warranted in such sample surveys.

In this chapter we have shown how, through quadrat spatial sampling, one can introduce stratifications and spatial controls that lead to an appropriate geographical spread and balance in the sample. By realizing an approximately proportional representation of the various components of the study area in the sample, we ensure that both the outlying and the central parts of the study are properly represented in the sample. The aim of this sampling strategy, as with most other sampling strategies, is to reduce the variance of unbiased estimates obtained at a given cost.

The principal conclusion of this chapter is that the placement of some rather simple spatial controls on simple random sampling can significantly reduce the variation of the resulting estimates of universe parameters. The greater geographical spread and balance that is achieved by such controls leads to more representative samples, and these in turn tend to produce more precise inferences about the attributes of the universe that are the concern of the sampling study.

Conclusion

Two fundamental ideas have achieved a more or less central position in theoretical research on intraurban spatial structure. These are the idea of order and the idea of causes acting to promote this order. This study has been concerned largely with a third idea which, since the turn of the century, has become increasingly important in scientific research—the idea of chance.

The many theoretical attempts to account for the spatial disposition of human activities in urban areas typically assume that such patterns have an identifiable order—an observable regularity which may be generalized to include cities of different size, type, and age. This spatial order is in a sense imposed on urban patterns by means of a classification system that groups the various activities into categories (not of identical activities but of activities which seem to behave alike), and then records the occurrence of these categories in urban space. The process of ordering is thus an empirical activity involving trial and error. Moreover the arrangement of activities into groups is predicated on attributes which are judged to be particularly relevant in terms of the kind of problem that is being studied. The only test of the 'correctness' of any particular order is its success in accounting for empirical observations; that is, a classification system cannot be judged in advance.

The concept of cause has been a central idea of science since the time of Newton. Thus it is not surprising to find its appearance in theoretical speculations on urban spatial structure The idea of cause in the context of the latter problem asserts that a definite configuration of factors acting together in a specified environment will always produce the same observable spatial pattern. The present state of the urban field determines all future states, and thus from a given beginning will always follow the same ends. In short, such a view identifies spatial structure as a machine which possesses definite properties that can be isolated and reproduced in space and time, and the behavior of which can be predicted.

The introduction of the concept of chance does not reject the notion of spatial structure as a machine, but merely asserts that more than one result may occur from the same beginning. Moreover it specifies that these results must occur in relatively fixed proportions over repeated trials. The idea of order is retained; however, its character now is probabilistic.

Order, cause, and chance were fused in this study to analyze urban retail spatial dispersion. Spatial order was imposed on the geographical arrangement of retail activities in urban areas by grouping them into classes, and by describing the spatial dispersion of each class by means of a probability function. Cause was introduced by a fundamental

Conclusion

hypothesis which held that due to profit considerations, shopping goods stores exhibited a mutual affinity, whereas convenience goods stores, in contrast, tended to repel one another. Finally, retail spatial dispersions were analyzed using stochastic models in an attempt to identify the particular chance mechanism that appeared to be producing the observed pattern of spatial dispersion.

More generally this study has endeavored to develop means whereby the spatial dispersion of various classes of human activities in urban areas can be quantitatively expressed, systematically analyzed, and compared on both an intraurban and an interurban level. The study proceeded by first translating certain fundamental hypotheses regarding clustering and dispersal into model form; it then deduced the spatial implications of these assumptions; and finally it statistically tested the significance of the match between the deduced implications and the empirical data.

It has been suggested that the paucity of empirical data on the physical dispositions of urban populations and activities is one of the principal impediments to the development of a viable theory of intraurban spatial structure (Jones, 1960). The need for appropriate measures with which to condense and summarize such data into comprehensible form is clear. The methods outlined in this study seem to have considerable merit in this respect. They may be used to answer questions concerning the temporal changes of a city's spatial dispersion. They suggest processes that may have generated this dispersion. Finally, they allow interurban comparisons to be made—a fundamental requisite in theory building.

References

Aitchison, J., Brown, J. A. C., 1957, *The Lognormal Distribution* (Cambridge University Press, Cambridge).

Allen, R. G. D., 1967, *Macro-Economic Theory: A Mathematical Treatment* (Macmillan, London), pp.49-52.

Arbous, A. G., Kerrich, J. E., 1951, "Accident statistics and the concept of accident-proneness", *Biometrics*, 7, 340-432.

Artle, R. K., 1965, *The Structure of the Stockholm Economy*, American edition (Cornell University Press, New York).

Aubert-Krier, J., 1954, "Monopolistic and imperfect competition in retail trade", in *Monopoly and Competition and Their Regulation*, Ed. E. H. Chamberlin (Macmillan, London), pp.281-300.

Bacharach, M., 1970, *Biproportional Matrices and Input-Output Change* (Cambridge University Press, London).

Barger, H., 1955, *Distribution's Place in the American Economy Since 1896* (Princeton University Press, Princeton, NJ).

Berry, B. J. L., 1967, *Geography of Market Centers and Retail Distribution* (Prentice-Hall, Englewood Cliffs, NJ).

Berry, B. J. L., Garrison, W. L., 1958a, "A note on central place theory and the range of a good", *Economic Geography*, 34, 304-311.

Berry, B. J. L., Garrison, W. L., 1958b, "Functional bases of the central place hierarchy", *Economic Geography*, 34, 145-154.

Berry, B. J. L., Pred, A., 1961, *Central Place Studies: A Bibliography of Theory and Applications* (Regional Science Research Institute, Philadelphia).

Bunge, W., 1962, *Theoretical Geography* (C. W. Gleerup, Lund, Sweden).

Chernoff, H., Lehmann, E. L., 1954, "The use of maximum likelihood estimates in X^2 tests for goodness of fit", *Annals of Mathematical Statistics*, 25, 579-586.

Clark, P. J., Evans, F. C., 1954a, "Distance to nearest neighbor as a measure of spatial relationships in populations", *Ecology*, 35, 445-453.

Cochran, W. G., 1952, "The X^2 test of goodness of fit", *Annals of Mathematical Statistics*, 23, 315-345.

Cochran, W. G., 1954, "Some methods for strengthening the common X^2 tests", *Biometrics*, 10, 417-451.

Curtis, J. T., McIntosh, R. P., 1950, "The interrelations of certain analytic and synthetic phytosociological characters", *Ecology*, 31, 434-455.

Edwards, C. B., Gurland, J., 1961, "A class of distributions applicable to accidents", *Journal of the American Statistical Association*, 56, 503-517.

Feller, W., 1957, *An Introduction to Probability Theory and Its Applications* (John Wiley, New York).

Fisher, R. A., 1924, "The conditions under which the chi square measures the discrepancy between observation and hypothesis", *Journal of the Royal Statistical Society*, 87, 442-450.

Fix, E., 1949, *Tables of the Noncentral χ^2* (University of California Press, Berkeley).

Freeman, H., 1963, *Introduction to Statistical Inference* (Addison-Wesley, Reading, Mass.).

Fuchs, V. R., 1968, *The Service Economy* (Columbia University Press, New York).

Goodman, R., Kish, L., 1950, "Controlled selection—a technique in probability sampling", *Journal of the American Statistical Association*, 45, 350-372.

Greig-Smith, P., 1952, "The use of random and contiguous quadrats in the study of the structure of plant communities", *Annals of Botany, London*, New Series, 16, 293-316.

Greig-Smith, P., 1964, *Quantitative Plant Ecology*, 2nd edition (Butterworths, London).

Gurland, J., 1957, "Some interrelations among compound and generalized distributions", *Biometrika*, 44, 265-268.

References

Hall, M., Knapp, J., Winsten, C., 1961, *Distribution in Great Britain and North America: A Study in Structure and Productivity* (Clarendon Press, Oxford).

Hart, P. E., Prais, S. J., 1956, "The analysis of business concentration", *Journal of the Royal Statistical Society,* Series A, part 2, **119**, 150-191.

Hinz, P., Gurland, J., 1970, "A test of fit for the negative binomial and other contagious distributions", *Journal of the American Statistical Association,* **65**, 887-903.

Holgate, P., 1964, "Estimation for the bivariate Poisson distribution", *Biometrika,* **51**, 241-245.

Holgate, P., 1966, "Bivariate generalizations of Neyman's Type A distribution", *Biometrika,* **53**, 241-244.

Hoover, E. M., 1948, *The Location of Economic Activity* (McGraw-Hill, New York).

Hotelling, H., 1929, "Stability in competition", *Economic Journal,* **39**, 41-57.

Hudson, J. C., Fowler, P. M., 1966, "The concept of pattern in geography", Discussion Paper Number 1, Department of Geography, University of Iowa, Iowa City.

Jessen, R. J., 1969, "Some methods of probability non-replacement sampling", *Journal of the American Statistical Association,* **64**, 175-193.

Jessen, R. J., 1970, "Probability sampling with marginal constraints", *Journal of the American Statistical Association,* **65**, 776-796.

Jones, B. G., 1960, *The Theory of the Urban Economy: Origins and Development with Emphasis on Intraurban Distribution of Population and Economic Activity,* Ph. D. Thesis, University of North Carolina, Chapel Hill, North Carolina.

Katti, S. K., Gurland, J., 1961, "The Poisson Pascal distribution", *Biometrics,* **17**, 527-538.

Katti, S. K., Gurland, J., 1962a, "Efficiency of certain methods of estimation for the negative binomial and the Neyman Type A distributions", *Biometrika,* **49**, 215-226.

Katti, S. K., Gurland, J., 1962b, "Some methods of estimation for the Poisson binomial distribution", *Biometrics,* **18**, 42-51.

Kendall, M. G., Stuart, A., 1961, *The Advanced Theory of Statistics, II: Inference and Relationship* (Charles Griffin, London).

Lewis, W. A., 1945, "Competition in retail trade", *Economica,* **12**, 202-234.

Lindgren, B. W., 1962, *Statistical Theory* (Macmillan, New York).

Lösch, A., 1954, *The Economics of Location,* translated by W. H. Woglom and W. F. Stolper (Yale University Press, New Haven, Conn.).

McAnally, A. P., 1963, "The measurement of productivity", *Journal of Industrial Economics,* **11**, 87-95.

McClelland, W. G., 1958, "Sales per person and size in retailing: some fallacies", *Journal of Industrial Economics,* **8**, 221-229.

McConnell, H., 1966, "Quadrat methods in map analysis", Discussion Paper Number 3, Department of Geography, University of Iowa, Iowa City.

Martin, J., Gomar, N., Rogers, A., 1969, "Some computer programs for quadrat analysis", WP, Center for Planning and Development Research, University of California, Berkeley.

Nelson, R. L., 1958, *The Selection of Retail Location* (F. W. Dodge, New York).

Neyman, J., 1939, "On a new class of 'contagious' distributions, applicable in entomology and bacteriology", *Annals of Mathematical Statistics,* **10**, 35-37.

Neyman, J., 1949, "Contribution to the theory of the χ^2 test", *Proceedings of the Berkeley Symposium on Mathematical Statistics and Probability* (University of California Press, Berkeley), pp.239-273.

Parzen, E., 1960, *Modern Probability Theory and Its Applications* (John Wiley, New York).

Patnaik, P. B., 1949, "The non-central χ^2- and F-distributions and their applications", *Biometrika*, **36**, 202-232.
Pennington, R. H., 1965, *Introductory Computer Methods and Numerical Analysis* (Macmillan, New York), pp.236-242.
Pielou, E. C., 1969, *An Introduction to Mathematical Ecology* (John Wiley, New York).
Quandt, R. E., 1964, "Statistical discrimination among alternative hypotheses and some economic regularities", *Journal of Regional Science*, **5**, 1-23.
Quandt, R. E., 1966, "Old and new methods of estimation and the Pareto distribution", *Metrika*, **10**, 55-82.
Rannells, J., 1956, *The Core of the City* (Columbia University Press, New York).
Ratcliff, R. U., 1939, *The Problem of Retail Site Selection* (Bureau of Business Research, School of Business Administration, University of Michigan, Ann Arbor, Michigan).
Rogers, A., Martin, J., 1971, "Quadrat analysis of urban dispersion: 3. Bivariate models", *Environment and Planning*, **3**, 433-450.
Roscoe, J. T., Byars, J. A., 1971, "An investigation of the restraints with respect to sample size commonly imposed on the use of the chi-square statistic", *Journal of the American Statistical Association*, **66**, 755-759.
Scott, P., 1970, *Geography and Retailing* (Aldine, Chicago).
Shenton, L. R., 1949, "On the efficiency of the method of moments and Neyman's Type A distribution", *Biometrika*, **36**, 450-454.
Shumway, R., Gurland, J., 1960b, "A fitting procedure for some generalized Poisson distributions", *Skandinavisk Aktuarietidskrift*, **43**, 87-108.
Sichel, H. S., 1951, "The estimation of the parameters of a negative binomial distribution with special reference to psychological data", *Psychometrika*, **16**, 107-127.
Simon, H. A., Bonini, C. P., 1958, "The size distribution of firms", *American Economic Review*, **48**, 607-617.
Skellam, J. G., 1952, "Studies in statistical ecology", *Biometrika*, **39**, 346-362.
Slakter, M. J., 1968, "Accuracy of an approximation to the power of the chi-square goodness of fit test with small but equal expected frequencies", *Journal of the American Statistical Association*, **63**, 912-918.
Southwood, T. R. E., 1966, *Ecological Methods* (Methuen, London).
Sprott, D. A., 1958, "The method of maximum likelihood applied to the Poisson binomial distribution", *Biometrika*, **14**, 97-106.
Stigler, G. J., 1956, *Trends in Employment in the Service Industries* (Princeton University Press, Princeton, NJ).
Teicher, H., 1954, "On the multivariate Poisson distribution", *Skandinavisk Aktuarietidskrift*, **37**, 1-9.
Thompson, H. R., 1956, "Distribution of distance to n-th neighbor in a population of randomly distributed individuals", *Ecology*, **37**, 391-394.
Thünen, J. H. von, 1826, *Der Isolierte Staat in Beziehung auf Landwirtschaft und Nationalökonomie* (Hamburg).
US Census, 1971, *Statistical Abstract of the United States: 1971 (92nd edition)* (Government Printing Office, Washington, DC).
Utton, M. A., 1970, *Industrial Concentration* (Penguin Modern Economics Series, Penguin Books, Baltimore, Maryland).
Weber, A., 1929, *Theory of the Location of Industries*, translated by C. J. Friedrich (The University of Chicago Press, Chicago).
Yarnold, J. K., 1970, "The minimum expectation in χ^2 goodness of fit tests and the accuracy of approximations for the null distribution", *Journal of the American Statistical Association*, **65**, 864-886.

List of relevant literature

Abrahamse, A. P. J., Van Der Laan, B. S., 1967, "On the power of three goodness-of-fit tests", Report 6713, Netherlands School of Economics, Econometric Institute, Rotterdam, unpublished manuscript, November 20, 1967.

Adelman, I., 1958, "A stochastic process of the size distribution of firms", *Journal of the American Statistical Association*, **53**, 893-904.

Adelson, R. M., 1966, "Compound Poisson distributions", *Operational Research Quarterly*, **17**, 73-75.

Anderson, T. W., Darling, D. A., 1952a, "A test of goodness of fit", *Journal of the American Statistical Association*, **49**, 765-769.

Anderson, T. W., Darling, D. A., 1952b, "Asymptotic theory of certain 'goodness of fit' criteria based on stochastic processes", *Annals of Mathematical Statistics*, **23**, 193-212.

Anscombe, F. J., 1948, "The transformation of Poisson, binomial and negative binomial data", *Biometrika*, **35**, 246-254.

Anscombe, F. J., 1949, "The statistical analysis of insect counts based on the negative binomial distribution", *Biometrics*, **5**, 165-173.

Anscombe, F. J., 1950, "Sampling theory of the negative-binomial and logarithmic series distributions", *Biometrika*, **37**, 358-382.

Archibald, E. E. A., 1948, "Plant populations. I. A new application of Neyman's contagious distribution", *Annals of Botany, London*, New Series, **12**, 221-235.

Archibald, E. E. A., 1950, "Plant populations. II. The estimation of the number of individuals per unit area of species in heterogeneous plant populations", *Annals of Botany, London*, New Series, **12**, 221-235.

Bates, G. E., Neyman, J., 1952, "Contributions to the theory of accident proneness. I. An optimistic model of the correlation between light and severe accidents", *University of California Publication in Statistics*, **1**, 215-253.

Beall, G., 1940, "The fit and significance of contagious distributions when applied to observations on larval insects", *Ecology*, **21**, 460-474.

Beall, G., Rescia, R. R., 1953, "A generalization of Neyman's contagious distributions", *Biometrics*, **8**, 334-386.

Bennett, B. M., 1959, "A sampling study on the power function of the X^2 'index of dispersion' test", *Journal of Hygiene*, **57**, 360-365.

Bennett, B. M., 1961, "A sampling study of the power function of the binomial X^2 'index of dispersion' test", *Journal of Hygiene*, **59**, 449-455.

Bennett, B. M., 1962, "On a heuristic treatment of the 'indices of dispersion'", *Annals of the Institute of Statistical Mathematics, Tokyo*, **14**, 151-157.

Berkson, J., 1938, "Some difficulties of interpretation encountered in the application of the chi-square test", *Journal of the American Statistical Association*, **33**, 526-536.

Berkson, J., 1940, "A note on the chi-square test, the Poisson and the binomial", *Journal of the American Statistical Association*, **35**, 362-367.

Bhattacharya, S. K., Holla, M. S., 1965, "On a discrete distribution with special reference to the theory of accident proneness", *Journal of the American Statistical Association*, **60**, 1060-1066.

Blackman, G. E., 1942, "Statistical and ecological studies in the distribution of species in plant communities. I. Dispersion as a factor in the study of changes in plant populations", *Annals of Botany, London*, New Series, **6**, 351-370.

Bliss, C. I., Fisher, R. A., 1953, "Fitting the negative binomial distribution to biological data, and note on the efficient fitting of the negative binomial", *Biometrics*, **9**, 176-200.

Bliss, C. I., Owen, A. R. G., 1958, "Negative binomial distributions with a common k", *Biometrika*, **45**, 37-58.

Bronowski, J., Neyman, J., 1945, "The variance of the measure of a two-dimensional random set", *Annals of Mathematical Statistics,* **16**, 330-341.
Camp, B. H., 1940, "Further comments on Berkson's problem", *Journal of the American Statistical Association,* **35**, 368-376.
Campbell, J. T., 1934, "The Poisson correlation function", *Proceedings of the Edinburgh Mathematical Society, Ser.2,* **4**, 18-26.
Clark, P. J., 1956, "Grouping in spatial distributions", *Science,* **123**, 373-374.
Clark, P. J., Evans, F. C., 1954b, "On some aspects of spatial pattern in biological populations", *Science,* **121**, 397-398.
Cochran, W. G., 1936, "The X^2 distribution for the binomial and Poisson series, with small expectations", *Annals of Eugenics,* **7**, 207-217.
Collins, N., Preston, L., 1961, "The size structure of the largest industrial firms 1909-1958", *American Economic Review,* **51**, 986-1011.
Cottam, G., Curtis, J. T., Hale, B. W., 1949, "A method for making rapid surveys of woodlands by means of pairs of randomly selected trees", *Ecology,* **30**, 101-104.
Dacey, M. F., 1962, "Analysis of central place and point patterns by a nearest neighbor method", *Proceedings of the International Geographical Union Symposium in Urban Geography, Lund 1960,* Ed. K. Norborg, *Lund Studies in Geography, B,* **24**, 55-75.
Dacey, M. F., 1963, "Order neighbor statistics for a class of random patterns in multidimensional space", *Annals of the Association of American Geographers,* **53**, 505-515.
Dacey, M. F., 1964, "Two-dimensional random point patterns: a review and an interpretation", *Papers of the Regional Science Association,* **13**, 41-55.
Dacey, M. F., 1965, "The geometry of central place theory", *Geografiska Annaler, B,* **47**, 111-124.
Dalenius, T., Hajek, J., Zubrzycki, S., 1961, "On plane sampling and related geometrical problems", *Proceedings of the Fourth Berkeley Symposium on Mathematics, Statistics, and Probability* (University of California Press, Berkeley), pp.125-150.
Daniels, H. E., 1961, "Mixtures of geometric distributions", *Journal of the Royal Statistical Society, Series B,* **23**, 409-413.
Darwin, J. H., 1957, "The power of the Poisson index of dispersion", *Biometrika,* **44**, 286-289.
Douglas, J. B., 1955, "Fitting the Neyman Type *A* (two parameter) contagious distribution", *Biometrics,* **11**, 149-173.
Dubey, S. D., 1966, "Compound Pascal distributions", *Annals of the Institute of Statistical Mathematics, Tokyo,* **18**, 357-365.
Evans, D. A., 1953, "Experimental evidence concerning contagious distributions in ecology", *Biometrika,* **40**, 186-211.
Feller, W., 1943, "On a general class of 'contagious' distributions", *Annals of Mathematical Statistics,* **14**, 389-400.
Fisher, R. A., 1941, "The negative binomial distribution", *Annals of Eugenics,* **11**, 182-187.
Fisher, R. A., Corbet, A. S., Williams, C. B., 1943, "The relation between the number of species and the number of individuals in a random sample from an animal population", *Journal of Animal Ecology,* **12**, 42-58.
Getis, A., 1964, "Temporal land use pattern analyses with the use of the nearest neighbor and quadrat methods", *Annals of the Association of American Geographers,* **54**, 391-398.
Goldman, M. I., 1962, "The cost and efficiency of distribution in the Soviet Union", *Quarterly Journal of Economics,* **70**, 437-453.

Goodall, D. W., 1952, "Quantitative aspects of plant distribution", *Biological Review,* **27**, 194-245.

Greenwood, H., Yule, G. U., 1920, "An inquiry into the nature of frequency distributions representative of multiple happenings", *Journal of the Royal Statistical Society,* Series A, **83**, 255-279.

Gurland, J., 1958, "A generalized class of contagious distributions", *Biometrics,* **14**, 229-249.

Gurland, J., 1959, "Some applications of the negative binomial and other contagious distributions", *American Journal of Public Health,* **49**, 1388-1399.

Haight, F. A., 1959, "The generalized Poisson distribution", *Annals of the Institute of Statistical Mathematics, Tokyo,* **11**, 101-105.

Haight, F. A., 1967, *Handbook of the Poisson Distribution* (John Wiley, New York).

Haldane, J. B. S., 1941, "The fitting of binomial distributions", *Annals of Eugenics,* **11**, 179-181.

Hall, M., Knapp, J., 1955, "Gross margins and efficiency measurement in retail trade", *Oxford Economic Papers,* **7**, 312-326.

Harvey, D. W., 1966, "Geographical processes and the analysis of point patterns", *Transactions and Papers of the Institute of British Geographers,* **40**, 81-95.

Hodges, J. L., Jr., Le Cam, L., 1960, "The Poisson approximation to the Poisson binomial distribution", *Annals of Mathematical Statistics,* **31**, 737-740.

Hoel, P. G., 1943, "On indices of dispersion", *Annals of Mathematical Statistics,* **14**, 155-162.

Holgate, P., 1965a, "Tests of randomness based on distance methods", *Biometrika,* **52**, 345-353.

Holgate, P., 1965b, "The distance from a random point to the nearest point of a closely packed lattice", *Biometrika,* **52**, 261-263.

Holla, M. S., Bhattacharya, S. K., 1964, "On a discrete compound distribution", *Annals of the Institute of Statistical Mathematics, Tokyo,* **16**, 377-384.

Ishii, G., Hayakawa, R., 1960, "On the compound binomial distribution", *Annals of the Institute of Statistical Mathematics, Tokyo,* **12**, 69-80.

Johnson, N. L., Kotz, S., 1969, *Discrete Distributions* (Houghton Mifflin, Boston).

Katti, S. K., 1960, *Some Aspects of Statistical Inference for Contagious Distributions,* Ph. D. Thesis, Iowa State University, Ames, Iowa.

Kemp, C. D., Kemp, A. W., 1956, "The analysis of point quadrat data", *Australian Journal of Botany,* **4**, 167-174.

Kemp, C. D., 1967, "On a contagious distribution suggested for accident data", *Biometrics,* **23**, 241-255.

Kendall, D. G., 1948, "On some modes of population growth leading to R. A. Fisher's logarithmic series distribution", *Biometrika,* **35**, 6-15.

Khatri, C. G., 1962, "A fitting procedure for a generalized binomial distribution", *Annals of the Institute of Statistical Mathematics, Tokyo,* **14**, 133-141.

Khatri, C. G., Patel, I. R., 1961, "Three classes of univariate discrete distributions", *Biometrics,* **17**, 567-575.

Krishnamoorthy, A. S., 1951, "Multivariate binomial and Poisson distribution", *Sankhya,* **11**, 117-124.

Lewis, P. A. W., 1965, "Some results on tests for Poisson processes", *Biometrika,* **52**, 67-77.

Liu, B. C., 1970, "Determinants of retail sales in large metropolitan areas, 1954 and 1963", *Journal of the American Statistical Association,* **65**, 1460-1473.

Maceda, E. C., 1948, "On the compound and generalized Poisson distributions", *Annals of Mathematical Statistics,* **19**, 414-416.

McGuire, J. U., Brindley, T. A., Bancroft, T. A., 1957, "The distribution of European corn borer larvae *pyrausta nubilalis* (*hbn*) in field corn", *Biometrics*, **13**, 65-78; **14**, 432-434 (errata and extensions).

Mann, H. B., Wald, A., 1942, "On the choice of the number of intervals in the application of the X^2 test", *Annals of Mathematical Statistics*, **13**, 306-317.

Maritz, J. S., 1950, "On the validity of inferences drawn from the fitting of Poisson and negative binomial distributions to observed accident data", *Psychological Bulletin*, **47**, 434-443.

Maritz, J. S., 1952, "Note on a certain family of discrete distributions", *Biometrika*, **39**, 196-198.

Martin, D. C., Katti, S. K., 1965, "Fitting of certain contagious distributions to some available data by the maximum likelihood method", *Biometrics*, **21**, 34-48.

Matuszewski, T. I., 1962, "Some properties of Pascal distribution for finite population", *Journal of the American Statistical Association*, **57**, 172-174.

Moore, P. J., 1954, "Spacing in plant populations", *Ecology*, **35**, 222-227.

Morisita, M., 1954, "Estimation of population density by spacing method", *Memoirs of the Faculty of Science, Kyushu University*, Series E, **1**, 187-197.

Naus, J. I., 1965, "Clustering of random points in two dimensions", *Biometrika*, **52**, 263-267.

Neyman, J., Scott, E. L., 1952, "A theory of the spatial distribution of galaxies", *Astrophysical Journal*, **116**, 144-163.

Neyman, J., Scott, E. L., 1957, "On a mathematical theory of populations conceived as conglomerations of clusters", *Cold Spring Harbour Symposia on Quantitative Biology*, **22**, 109-120.

Neyman, J., Scott, E. L., 1959, "Stochastic models of population dynamics", *Science*, **130**, 303-308.

O'Carroll, F. M., 1962, "Fitting a negative binomial distribution to coarsely grouped data by maximum likelihood", *Applied Statistics*, **11**, 196-201.

Ottestad, P., 1944, "On certain compound frequency distributions", *Skandinavisk Aktuarietidskrift*, **27**, 32-42.

Patil, G. P., 1960, "On the evaluation of the negative binomial distribution with examples", *Technometrics*, **2**, 501-505.

Patil, G. P., 1962a, "Maximum likelihood estimation for generalized power series distributions and its application to a truncated binomial distribution", *Biometrika*, **49**, 227-238.

Patil, G. P., 1962b, "Certain properties of the generalized power series distribution", *Annals of the Institute of Statistical Mathematics, Tokyo*, **14**, 179-182.

Patil, G. P., 1964, "On certain compound Poisson and compound binomial distributions", *Sankhya, A*, **26**, 293-294.

Patil, G. P. (Ed.), 1965a, *Classical and Contagious Discrete Distributions* (Pergamon Press, Oxford).

Patil, G. P., 1965b, "Certain characteristic properties of multivariate discrete probability distributions akin to the Bates-Neyman model in the theory of accident proneness", *Sankhya, A*, **27**, 259-270.

Patil, G. P., Bildikar, S., 1967, "Multivariate logarithmic series distribution as a probability model in population and community ecology and some of its statistical properties", *Journal of the American Statistical Association*, **62**, 655-674.

Pielou, E. C., 1957, "The effect of quadrat size on the estimation of the parameters of Neyman's and Thomas' distributions", *Journal of Ecology*, **45**, 31-47.

Pielou, E. C., 1960, "A single mechanism to account for regular, random and aggregated populations", *Journal of Ecology*, **48**, 575-584.

Pollard, S., Hughes, J. D., 1955, "Retailing costs: some comments on the census of distribution, 1950", *Oxford Economic Papers*, **7**, 71-93.

Quenouille, M. H., 1949, "A relation between the logarithmic, Poisson, and negative binomial series", *Biometrics*, **5**, 162-164.
Robinson, P., 1954, "The distribution of plant populations", *Annals of Botany, London*, New Series, **18**, 35-45.
Rogers, A., 1965, "A stochastic analysis of the spatial clustering of retail establishments", *Journal of the American Statistical Association*, **60**, 1094-1103.
Rogers, A., 1969a, "Quadrat analysis of urban dispersion: 1. Theoretical techniques", *Environment and Planning*, **1**, 47-80.
Rogers, A., 1969b, "Quadrat analysis of urban dispersion: 2. Case studies of urban retail systems", *Environment and Planning*, **1**, 155-171.
Rogers, A., Gomar, N., 1969, "Statistical inference in quadrat analysis", *Geographical Analysis*, **1**, 370-384.
Rogers, A., Raquillet, R., 1972, "Quadrat analysis of urban dispersion: 4. Spatial sampling", *Environment and Planning*, **4**, 331-345.
Satterthwaite, F. E., 1942, "Generalized Poisson distribution", *Annals of Mathematical Statistics*, **13**, 410-417.
Schilling, W., 1947, "A frequency distribution represented as the sum of two Poisson distributions", *Journal of the American Statistical Association*, **42**, 407-424.
Shenton, L. R., 1950, "Maximum likelihood and the efficiency of the method of moments", *Biometrika*, **37**, 111-116.
Shenton, L. R., 1958, "Moment estimators and maximum likelihood", *Biometrika*, **45**, 411-420.
Shenton, L. R., 1959, "The distribution of moment estimators", *Biometrika*, **46**, 296-305.
Shenton, L. R., Wallington, P. A., 1962, "The bias of moment estimators with an application to the negative binomial distribution", *Biometrika*, **49**, 193-204.
Shepherd, W. G., 1961, "A comparison of industrial concentration in the United States and Britain", *Review of Economics and Statistics*, **43**, 70-75.
Shumway, R., Gurland, J., 1960a, "Fitting the Poisson binomial distribution", *Biometrics*, **16**, 522-533.
Sibuya, M., Yoshimura, I., Shimizu, R., 1964, "Negative multinomial distribution", *Annals of the Institute of Statistical Mathematics*, **16**, 409-426.
Silberman, I. H., 1967, "On lognormality as a summary measure of concentration", *American Economic Review*, **107**, 807-830.
Simon, H. A., 1955, "On a class of skew distribution functions", *Biometrika*, **42**, 425-440.
Singer, E. M., 1968, *Antitrust Economics* (Prentice-Hall, Englewood Cliffs, NJ).
Skellam, J. G., 1948, "A probability distribution derived from the binomial distribution by regarding the probability of success as variable between the sets of trials", *Journal of the Royal Statistical Society*, Series B, **10B**, 257-261.
Skellam, J. G., 1958, "On the derivation and applicability of Neyman's Type A distribution", *Biometrika*, **45**, 32-36.
Somerville, P. N., 1957, "Optimum sampling in binomial populations", *Journal of the American Statistical Association*, **52**, 494-502.
Subrahmaniam, K., 1965, "On a general class of contagious distributions: the Pascal-Poisson distribution", Paper Number 356, Department of Biostatistics, Johns Hopkins University, Baltimore, Maryland.
Sukhatme, P. V., 1938, "On the distribution of X^2 in samples of the Poisson series", *Journal of the Royal Statistical Society Supplement*, **1**, 75-79.
Teicher, H., 1960, "On the mixture of distribution", *Annals of Mathematical Statistics*, **31**, 55-73.
Thomas, M., 1949, "A generalization of Poisson's binomial limit for use in ecology", *Biometrika*, **36**, 18-25.

Thompson, H. R., 1954, "A note on contagious distributions", *Biometrika,* **41,** 268-271.
Thompson, H. R., 1955, "Spatial point processes with application to ecology", *Biometrika,* **42,** 102-115.
Thompson, H. R., 1958, "The statistical study of plant distribution patterns using a grid of quadrats", *Australian Journal of Botany,* **6,** 322-343.
Torii, T., 1956, *The Stochastic Approach in Field Population Ecology* (Japan Society for the Promotion of Science, Tokyo).
Tweedie, M. C. K., 1952, "The estimation of parameters from sequentially sampled data on a discrete distribution", *Journal of the Royal Statistical Society,* Series B, **14,** 238-245.
Walsh, J. E., 1963, "Bounded probability properties of Kolmogorov-Smirnov and similar statistics for discrete data", *Annals of the Institute of Statistical Mathematics,* **15,** 153-158.
Williams, C. A., Jr., 1950, "On the choice of the number and width of classes for the chi-square test of goodness of fit", *Journal of the American Statistical Association,* **45,** 77-86.
Winsten, C., Hall, M., 1961, "The measurement of economies of scale", *Journal of Industrial Economics,* **9,** 255-264.

Author index

Aitchison, J., 82
Allen, R. G. D., 79
Arbous, A. G., 12, 126
Artle, R. K., 92, 102, 103, 105
Aubert-Krier, J., 73

Bacharach, M., 146
Barger, H., 71
Berry, B. J. L., 74, 75, 88
Bonini, C. P., 82
Brown, J. A. C., 82
Bunge, W., 1
Byars, J. A., 68

Chernoff, H., 31, 68
Clark, P. J., 8, 9
Cochran, W. G., 57, 68
Curtis, J. T., 108

Edwards, C. B., 123
Evans, F. C., 8, 9

Feller, W., Preface, 13, 22
Fisher, R. A., 68
Fix, E., 58
Fowler, P. M., 1
Freeman, H., Preface
Fuchs, V. R., 71

Garrison, W. L., 88
Gomar, N., 18
Goodman, R., 132
Greig-Smith, P., 5, 6, 105, 108
Gurland, J., 21, 27, 28, 30, 31, 37, 38, 41, 42, 45, 46, 49, 51, 57, 123

Hall, M., 71, 75
Hart, P. E., 80, 82
Hinz, P., 57
Holgate, P., 118, 124, 125
Hoover, E. M., 98
Hotelling, H., 88
Hudson, J. C., 1

Jessen, R. J., 132, 133, 134
Jones, B. G., 151

Katti, S. K., 28, 30, 31, 37, 38, 41, 42, 45, 46, 51
Kendall, M. G., 56, 68
Kerrich, J. C., 12, 126
Kish, L., 132
Knapp, J., 71, 75

Lehmann, E. L., 31, 68
Lewis, W. A., 73
Lindgren, B. W., Preface
Lösch, A., 74

McAnally, A. P., 79
McClelland, W. G., 75
McConnell, H., 2
McIntosh, R. P., 108
Martin, J., 18, 126

Nelson, R. L., 88, 116
Neyman, J., 24, 27, 57

Parzen, E., Preface, 23
Patnaik, P. B., 58, 59, 70
Pennington, R. H., 37
Pielou, E. C., 97
Prais, S. J., 80, 82
Pred, A., 75

Quandt, R. E., 82

Rannells, J., 87
Ratcliff, R. U., 88
Rogers, A., 18, 126
Roscoe, J. T., 68

Scott, P., 74
Shenton, L. R., 41, 42
Shumway, R., 49
Sichel, H. S., 37, 38
Simon, H. A., 82
Skellam, J. G., 27, 28
Slakter, M. J., 68
Southwood, T. R. E., 5
Sprott, D. A., 45, 46
Stigler, G. J., 71
Stuart, A., 56, 68

Teicher, H., 118
Thompson, H. R., 10
Thünen, J. H. von, 71

Utton, M. A., 80

Weber, A., 71
Winsten, C., 71, 75

Yarnold, J. K., 68

Subject index

Bias ratio 140
Binomial distribution 15-16, 30, 118
Binomial distribution,
   bivariate model of 118
   maximum likelihood estimator for 34-35
   mean of 15
   moment estimator for 34
   probability density function of 15
   probability generating function of 15
   simulation of 18-20
   variance-mean ratio of 16
   variance of 15

Chi-square goodness of fit test 7, 54-57
Chi-square goodness of fit test,
   estimation problems 68
   grouping problems 68-70
   power of the test 57-67
Chi-square statistic 7
Clustered point process 5
Clustered spatial point pattern 5, 20
Clustered spatial point pattern,
   perfect clustering in 5
Coefficient of variation 141
Compound distribution 21, 122
Compound distribution,
   binomial 27
   bivariate correlated Poisson 123
   negative binomial 28
   Poisson 23
Concentration ratio 80

Dispersion 1 (see Spatial dispersion)
Dispersion line 97, 115

Estimation (see Parameter estimation)

Fundamental component distributions 13, 120

Gamma distribution 24, 123
Generalized distribution 21, 124
Generalized distribution,
   bivariate correlated Poisson 125
   Poisson 25-26

Hypothesis testing 54

Ljubljana, Yugoslavia 75-78, 90-100, 112-115, 140-142
Logarithmic distribution 25
Lognormal distribution 82
Lorenz curve 81
Lorenz index of concentration 84

Maximum likelihood estimators 33, 130-131
Moment estimators 31-32, 126, 129
Multinomial distribution 55

Nearest neighbor analysis 8-11, 97
Negative binomial distribution 16-18, 24-25, 30, 59, 62, 70, 120-124
Negative binomial distribution,
   bivariate models of 120-124, 126, 129-131
   existence and efficiency of estimators for 37-38
   maximum likelihood estimators for 36-37
   mean of 17
   moment estimators for 35
   probability density function of 17-18
   probability generating function of 16, 18
   simulation of 18-20
   variance of 17
   variance-mean ratio of 17
Newton-Raphson method 37, 40, 44, 51
Neyman Type A distribution 23-25, 30, 59, 62, 70, 124-126
Neyman Type A distribution,
   bivariate model of 124-126, 129-131
   existence and efficiency of estimators for 41-42
   maximum likelihood estimators for 39-41
   mean of 24
   moment estimators for 39
   probability density function of 24
   probability generating function of 23
   variance of 24
Noncentral chi-square distribution 57-58

Subject index

Parameter estimation 31-34, 126-131
Parameter estimators,
  maximum likelihood 33, 130-131
  moment 31-32, 126, 129
  ratio of first two observed frequencies 51
  sample zero frequency 45-46
Pareto distribution 82
Pattern 1 (see also Regular, Random, and Clustered spatial point patterns)
Poisson distribution 4, 7, 8, 13-14, 30, 116-120
Poisson distribution,
  bivariate model of 116-120, 126
  maximum likelihood estimator for 34
  mean of 4
  moment estimator for 33
  probability density function of 4
  probability generating function of 14
  simulation of 18-20
  variance of 4
  variance-mean ratio of 14
Poisson-binomial distribution 27, 30
Poisson-binomial distribution,
  existence and efficiency of estimators for 45-47
  maximum likelihood estimators for 43-45
  mean of 27
  moment estimators for 42-43
  probability density function of 27
  probability generating function of 27
  sample zero frequency estimators for 45-47
  variance of 27
Poisson-negative binomial distribution 27-28, 30
Poisson-negative binomial distribution,
  existence and efficiency of estimators for 50-51
  maximum likelihood estimators for 49-50
  mean of 28
  moment estimators for 47-48
  probability density function of 28
  probability generating function of 28
  ratio of first two observed frequencies estimators for 51
  variance of 28

Population spatial dispersion 98-100, 126-131
Production function 78-79

Quadrat analysis 5-8, 13, 52, 116
Quadrat analysis,
  bivariate models of 116
  fundamental component distributions of 13
  optimal quadrat size in 105-111
  quadrat censusing in 52
  quadrat sampling in 52
Quadrat, definition of 5
Quadrat size 8, 105-111
Quadrat size,
  optimal 105-111
  problem of 8
Quadrat spatial sampling 138-140
Quadrat spatial sampling,
  bias ratio, definition of 140
  coefficient of variation, definition of 141
  definition of 138-140
  sensitivity tests of 142-145
  simulation of 140
  spatially biased selection in 145
Quadrat spatial sampling
  in Ljubljana, Yugoslavia 140-142
  with estimated constraints 145-148

Randomness 2
Random point process 5
Random spatial point pattern 2, 5, 8
Regular point process 5
Regular spatial point pattern 5, 20
Regular spatial point pattern,
  perfect regularity in 5
Retail location 74-75
Retail spatial dispersion 88-89
Retail spatial dispersion,
  compound model of 89
  dispersion line for 97, 115
  generalized model of 89-90
Retail spatial dispersion
  in Ljubljana, Yugoslavia 90-100, 112-115
  in San Francisco, USA 100-102
  in Stockholm, Sweden 102-105
  of employment, space, and sales 112-115

Retailing 71-75
Retailing,
    classes of goods in 72-73
    competition in 73
    location in 74-75
    productivity and size in 78-80
    size concentration of 80-85
    spatial concentration of 85-87
Retailing in Ljubljana, Yugoslavia
    75-78

Sample zero frequency estimators
    45-46
Sampling 132
Sampling,
    Jessen algorithm for 136-138
    Jessen method of 133-136
    quadrat spatial 141
    simple random 141
    spatial 138
    stratified random 132
Sampling with marginal constraints 133
San Francisco, USA 100-102
Shape 1

Simulation 18-20, 61-67, 140
Simulation
    of quadrat spatial sampling 140
    of the chi-square power function
        61-67
    of the fundamental component
        distributions 18-20
Spatial affinity 96
Spatial dispersion 1-2, 12-13,
    116-126 (see also Retail spatial
    dispersion)
Spatial dispersion,
    bivariate models of 116-126
    clustered 13, 16-18
    random 13-14
    random bivariate 116-120
    regular 13, 15-16
    statistical analysis of 1-5
Stockholm, Sweden 102-105

Type I and II errors 61

Variance-mean ratio 6